T0313818

Advanced Technologies and Wireless Networks Beyond 4G

Advanced Technologies and Wireless Networks Beyond 4G

Nathan Blaunstein
Ben Gurion University of the Negev
Beer Sheva, IS, 74105

Dr. Yehuda Ben-Shimol
Ben Gurion University of the Negev
Shay Agnon 4/27, Beer Sheva, IS, 84758

This edition first published 2021
© 2021 John Wiley & Sons, Inc.

The right of Nathan Blaunstein and Dr. Yehuda Ben-Shimol to be identified as the authors of this work has been asserted in accordance with law.

Registered Office
John Wiley & Sons, Inc., 111 River Street, Hoboken, NJ 07030, USA

Editorial Office
John Wiley & Sons, Inc., 111 River Street, Hoboken, NJ 07030, USA

For details of our global editorial offices, customer services, and more information about Wiley products visit us at www.wiley.com.

Wiley also publishes its books in a variety of electronic formats and by print-on-demand. Some content that appears in standard print versions of this book may not be available in other formats.

Library of Congress Cataloging-in-Publication Data

Names: Blaunstein, Nathan, author. | Ben-Shimol, Yehuda, author.
Title: Advanced technologies and wireless networks beyond 4G / Nathan
    Blaunstein, Ben Gurion University of the Negev, Beer Sheva, IS, Dr.
    Yehuda Ben-Shimol, Ben Gurion University of the Negev, Beer Sheva, IS.
Description: Hoboken, NJ, USA : John Wiley & Sons, Inc., 2021. | Includes
    bibliographical references and index.
Identifiers: LCCN 2020024256 (print) | LCCN 2020024257 (ebook) | ISBN
    9781119692447 (cloth) | ISBN 9781119692409 (adobe pdf) | ISBN
    9781119692454 (epub)
Subjects: LCSH: Wireless communication systems.
Classification: LCC TK5103.2 .B555 2021 (print) | LCC TK5103.2 (ebook) |
    DDC 621.384–dc23
LC record available at https://lccn.loc.gov/2020024256
LC ebook record available at https://lccn.loc.gov/2020024257

Cover design by Wiley
Cover image: © zf L/Getty Images

Set in 10/12pt WarnockPro by SPi Global, Chennai, India

10 9 8 7 6 5 4 3 2 1

Contents

Acknowledgements

It is with the kind permission of Yehuda Ben-Shimol, my coauthor, that I write the acknowledgments for this book.

This book would never see the light of day without the encouragement and support of our colleagues. First of all, we thank Prof. Christos Christodoulou, our coauthor of two of our previous books, also published by Wiley & Sons in 2007 and 2017. These books stimulated both of us to finalize our vision on how to predict operational parameters of any terrestrial wireless network, from 2G to 5G. This is a subject that was outside the matter of the two previous books that dealt with radio propagation and adaptive antennas for various terrestrial, atmospheric, and ionospheric environments. We are grateful to him and to those numerous colleagues who encouraged us and influenced our thoughts and ideas.

We owe special thanks to the former MSc and PhD students, Fred Tsimet, Natalie Yarkoni, Dmitry Katz, and Evgeny Tsalolikhin. Their helpful contributions were in modeling and computing in the areas of data stream capacity prediction in novel wireless networks, more specifically, interuser interference (IUI) and intersymbol interference (ISI) in various urban environments with dense buildings' layout and the huge subscribers' density, and for estimation of users' capacity in complicated hierarchy of femto/pico/micro/macrocell arrangement.

We appreciate the hard work of Amudhapriya Sivamurthy, editorial staff of Wiley & Sons publisher group and their reviewers who transformed our drawings and graphs into elegant and attractive figures and who did their best to present the final text with clarity and precision.

Finally, we are greatly indebted to our families for providing the kind atmosphere in which this book "saw the light."

Preface

Nowadays, hundreds of millions of people around the world are involved in cellular and noncellular wireless communications. It is purely a matter of convenience: receive and make audio and video calls at your leisure, send and receive SMS, MMS, etc., surf the WEB and send e-mails, or reach formats of online messages, anytime and almost everywhere. The mobile phone has become a fashionable and everyday object. In the present "information age," various subscribers find the necessity to access data while on the move or need to be connected 24 hours a day, 7 days a week to the outside world. Moreover, subscribers located at helicopters and aircrafts also want to be serviced any time and in any place from ground-based radio networks with a good quality of service (QoS) and grade of service (GoS).

During the past century, the cellular network has gone through three generations (from 1G to 3G). The first generation (1G) of cellular networks was analog in nature. To accommodate more cellular phone subscribers, digital TDMA (time-division multiple access), FDMA (frequency-division multiple access), and CDMA (code-division multiple access) technologies were used in the second generation (2G), to increase the network capacity. With digital technologies, digitized voice can be coded and encrypted. Therefore, the 2G cellular network is also more secure.

The third generation (3G) integrates cellular phones into the internet by providing high-speed packet-switching data transmission in addition to circuit-switching voice transmission. The 3G cellular networks have been deployed in some parts of Asia, Europe, and the United States since 2002, and are now widely deployed all over the world (full information on the matter can be seen in the books [1–12] and in bibliography therein).

By 2009, it had become clear that, at some point, 3G networks would be overwhelmed by the growth of bandwidth-intensive applications like streaming media. Consequently, the industry began looking for data-optimized technologies, with the promise of speed, capacity, and data rates improvements, with emphasis on GoS and C/I (carrier to interference ratio), especially in urban or suburban environments, with different density and rates of calls and/or data flow which can change at any time [13, 14].

Moreover, the demand for mobile data access is intense and will continue to increase exponentially in the foreseeable future. The only clear way to increase the capacity by the orders of magnitude required over the recent and the next decade is by adding more network infrastructure. These trends require fundamentally new network approaches to deploying such infrastructure in a cost-effective manner [13, 14].

A key recent trend in this regard is the use of femtocells overlaid throughout the traditional tower-based network. These small, inexpensive, and short-range access points can be deployed either by the end user or by the service provider and typically occupy licensed spectrum and have an IP backhaul [15–18].

There are several issues in the macrocell networks, especially in urban or suburban areas with medium or high density of cellular users. When mobile users are within a range close to the macro antenna, the C/I is about 20 dB, which means good capacity and good data rates of the channel. However, when mobile users are approaching toward the edges of a cell, especially inside buildings, the C/I gets reduced (about 3 dB), which means that users are more susceptible to interference and obtain very low data rate [19]. To solve this problem, communication operators went to consult new technologies, such as femtocell, to improve the capacity and data rates of the channel [20–22, 11, 12, 23–29].

This is why the fourth-generation (4G) cellular networks are based on the advanced technologies and the so-called adaptive/smart antennas, combined with multiple-input-multiple-output (MIMO) configurations. These, combined with cellular planning strategy, from macrocell to femtocell networks, are needed to satisfy the increasing demand for the implementation of modern terrestrial wireless systems [1–5, 13].

Therefore, this book, as a continuation and extension of the two books published in 2007 and 2014, respectively (see [11, 12]), deals now with new wireless networks, from fourth generation (4G) to fifth generation (5G), in their historical perspective, and accounting for the corresponding so-called physical layer of each of them, such as multicarrier MIMO techniques, orthogonal frequency-division multiplex (OFDM), orthogonal frequency-division multiple access (OFDMA), LTE-MIMO, and other advanced technologies.

This book has been arranged to account for several current and modern wireless networks and the corresponding novel technologies and techniques based on the main aspects of "physical layer" — radio propagation phenomena in various terrestrial wireless communication links. The practical aspects of the presented approach can be easily transferred to atmospheric (aircraft) and satellite communication links. These aspects have been described in detail in [11, 12]. In [12], the main aspects of cellular and noncellular multiple-access networks based on standard technologies, such as frequency-division multiple access (FDMA), time-division multiple access (TDMA), and space-division multiple access (SDMA), which were mostly created for 3G network operational characteristics prediction, such as GoS, QoS, data stream capacity, and spectral efficiency passing such types of wireless networks, are briefly described.

As mentioned above, during the recent decades new advanced technologies were presented, such as MIMO, based on the usage of multibeam adaptive antennas, OFDM and OFDMA multiple-access techniques, and multicarrier service based on the integration of femto/pico/micro/macrocell concepts of cellular layout deployments. Therefore, this book is intended to appeal to any scientist, practical engineer, or designer, who is concerned with the operation and service of various radio networks, current and modern, including personal, mobile, aircraft, and satellite communication.

The book is composed of four parts, consisting 10 chapters. Part I includes two chapters. Chapter 1 briefly describes the well-known networks–from 2G to 3G – in a historical perspective. In Chapter 2, the corresponding 2G to 3G technologies and networks are presented in brief, describing their advantages and disadvantages.

Part II consists of three chapters. Chapter 3 illuminates the so-called physical layer of the network, based on those propagation models in terrestrial communication links which have been found as more attractive to the real land communication through their comparison with the recent experiments carried out in the built-up terrain. Chapter 4 presents the polarization diversity analysis for networks beyond 4G and gives comparative analysis of depolarization effects and the corresponding path loss factor. Various land areas are considered: rural, mixed residential, suburban, and urban, based on the results obtained from the corresponding stochastic multiparametric approach as a combination of physical "layer" based on the propagation phenomena of radio waves in such kinds of the terrain and the statistical description of the terrain features. Chapter 5 discusses, via the same stochastic multiparametric approach, the positioning of any subscriber located in the areas of service, both for land-to-land and land-to-atmosphere communication links. Positioning is becoming actual with the development of networks of fourth and fifth generations.

Part III consists of one chapter. Chapter 6 illuminates the techniques of how to plan different integrated femto/pico/micro/macrocell deployments for increasing of the GoS of the multiuser 4G and 5G networks. Part IV consists of three chapters. In Chapter 7, the reader will meet with new technologies of time and frequency dispersion for the currently used 4G networks and the upcoming 5G new networks. In Chapter 8, MIMO modern network design in space and time domains is presented along with different techniques of signal data capacity and spectral efficiency based on the universal stochastic approach. Then, in Chapter 9, a MIMO network based on multibeam adaptive antennas integrated with modern LTE releases is discussed showing that the usage of MIMO system with adaptive multibeam antennas can significantly increase the capacity of signal data transmitting via LTE networks and users' layout in the areas of service, that is, increase grade of service and quality of service of modern networks beyond 4G. Part V consists of only one chapter–Chapter 10 – which deals with mega-cell satellite networks for land-to-satellite and satellite-to-land wireless communications. Here, the most adaptive and experimentally proved models are presented as a physical–statistical basis of any current and future satellite–land network. Then are presented the currently used networks of 3G to 4G and a brief information on new advanced networks. The main goal of this section is to present approaches on how to mitigate fading phenomena caused by the influence of terrain overlay profile (and first of all built-up overlay profile) on fading phenomena that, as shown in previous chapters, is a most dangerous source corrupting information and decreasing GOS and QoS in future satellite–land networks beyond 4G.

Let us now enter deeper into the subject, including the definition of the "physical layer" of each network on the basis of more attractive propagation models above the built-up terrain, "reaction" of each communication network on radio propagation, and the corresponding fading effects, as a main source of multiplicative noises, occurring in wireless communication networks that affect the signal data streams passing such kind of channels, leading to the significant loss of information – digital, analogue, video, audio, and so forth.

Nathan Blaunstein

Beer-Sheva, January 2020

References

1 Jakes, W.C. (1974). *Microwave Mobile Communications*. New York: Wiley.

2 Lee, S.C.Y. (1989). *Mobile Cellular Telecommunication Systems*. New York: McGraw-Hill.

3 Steele, R. (1992). *Mobile Radio Communication*. IEEE Press.

4 Proakis, J.G. (1995). *Digital Communications*, 3e. New York: McGraw-Hill.

5 Stuber, G.L. (1996). *Principles of Mobile Communications*. Boston, MA: Kluwer Academic Publishers.

6 Steele, R. and Hanzo, L. (1999). *Mobile Communications*, 2e. Chichester: Wiley.

7 Li, J.S. and Miller, L.E. (1998). *CDMA Systems Engineering Handbook*. Boston, MA and London: Artech House.

8 Saunders, S.R. (2001). *Antennas and Propagation for Wireless Communication Systems*. Chichester: Wiley.

9 Burr, A. (2001). *Modulation and Coding for Wireless Communications*. New York: Prentice Hall PTR.

10 Molisch, A.F. (ed.) (2000). *Wideband Wireless Digital Communications*. Chichester, London: Prentice Hall PTR.

11 Blaunstein, N. and Christodoulou, C. (2007). *Radio Propagation and Adaptive Antennas for Wireless Communication Links*, 1e. Hoboken, NJ: Wiley.

12 Blaunstein, N. and Christodoulou, C. (2014). *Radio Propagation and Adaptive Antennas for Wireless Communication Networks-Terrestrial, Atmospheric and Ionospheric*, 2e. Hoboken, NJ: Wiley.

13 Peterson, R.L., Ziemer, R.E., and Borth, D.E. (1995). *Introduction to Spread Spectrum Communications*. New York: Prentice Hall PTR.

14 Rappaport, T.S. (1996). *Wireless Communications: Principles and Practice*, 2e in 2001. New York: Prentice Hall PTR.

15 Paetzold, M. (2002). *Mobile Fading Channels: Modeling, Analysis, and Simulation*. Chichester: Wiley.

16 Simon, M.K., Omura, J.K., Scholtz, R.A., and Levitt, B.K. (1994). *Spread Spectrum Communications Handbook*. New York: McGraw-Hill.

17 Glisic, S. and Vucetic, B. (1997). *Spread Spectrum CDMA Systems for Wireless Communications*. Boston, MA and London: Artech House.

18 Dixon, R.C. (1994). *Spread Spectrum Systems with Commercial Applications*. Chichester: Wiley.

19 Viterbi, A.J. (1995). *CDMA: Principles of Spread Spectrum Communication, Addison-Wesley Wireless Communications Series*. Reading, MA: Addison-Wesley.

20 Goodman, D.J. (1997). *Wireless Personal Communication Systems*. Reading, MA: Addison-Wesley.

21 Schiller, J. (2003). *Mobile Communications, Addison-Wesley Wireless Communications Series*, 2e. Reading, MA: Addison-Wesley.

22 Molisch, A.F. (2007). *Wireless Communications*. Chichester: Wiley.

23 Hadar, O., Bronfman, I., and Blaunstein, N. (2017). Optimization of error concealment based on analysis of fading types, Part 1: Statistical Description and error concealment of video signals. *J. Inf. Control Syst.* 86 (1): 72–82.

24 Hadar, O., Bronfman, I., and Blaunstein, N. (2017). Optimization of error concealment based on analysis of fading types, Part 2: Modified and new models of video signal error concealment. Practical simulations and their results. *J. Inf. Control Syst.*, 86 (2): 67–76.

25 Sun, M.T. and Reibman, A.R. (2001). *Compressed Video over Networks*. New York: Marcel Dekker.

26 Doshkov, D., Ndjiki-Nya, P., Lakshman, H. et al. (2010). Towards efficient intra prediction based on image inpainting methods. *28th Picture Coding Symposium*, PCS2010, Nagoya, Japan (8–10 December 2010), 6 pages.

27 Chen, B.N. and Lin, Y. (2006). Selective motion field interpolation for temporal error concealment. *International Conference on Computer and Communication Engineering* (ICCCE 2006), Kuala Lumpur, Malaysia (9–11 May 2006).

28 Hadar, O., Huber, M., Huber, R., and Greenberg, S. (2005). New hybrid error concealment for digital compressed video. *EURASIP J. Appl. Signal Process.* 2005 (12): 1821–1833.

29 Blaunstein, N., Arnon, S., Zilberman, A., and Kopeika, N. (2010). *Applied Aspects of Optical Communication and LIDAR.* Boca Raton, FL: Taylor and Francis Group.

Acronyms

1G	first generation
2D	second generation
3D	third generation
4G	fourth generation
5G	fifth generation
AFD	average fade duration
AOA	angle of arrival
BER	bit error rate
BS	base station
CAPEX	capital expense
CCDF	complementary cumulative distribution function
CDF	cumulative distribution function
CDMA	code-division multiple access
C/I	carrier-to-interference ratio
CSG	closed subscriber group
DFT	direct Fourier transform
DS	Doppler shift
DSA	dedicated spectral assignment
EOA	elevation of arrival
FAP	femtocell access point
FDD	frequency-division duplexing
FDMA	frequency-division multiple access
FMC	femto-macro/microcellular (interference)
GEO	geostationary orbit (satellite)
GoS	grade of service
HAP	home access point
HSPA	high-speed packet access
IA	interference aware
IFT	inverse Fourier transform
ISE	individual subscriber element
ISI	intersymbol interference

IUI	interuser interference
LEO	low orbit (satellite)
LCR	level-crossing rate
LOS	line of sight
LSC	land–satellite communication
LTE	long-term evolution
MDM	minimum-distance measure
MDM-Hellinger	minimum-distance measure by use of Hellinger technique
MDM-KLD	minimum-distance measure by use of special-likelihood distance technique
MEO	medium orbit (satellite)
MIMO	multiple-input-multiple-output
MISO	multiple-input-single-output
ML	maximum likelihood (function)
MS	mobile subscriber
MU	mobile user
NLOS	non-line-of-sight
OFDM	orthonormal frequency-division multiplexing
OFDMA	orthonormal frequency-division multiplexing access
OSG	open subscriber group
PDF	probability density function
QoS	quality of service
RMS	root mean square
SC	single carrier
SF	system function
SIMO	single-input-multiple-output
SISO	single-input-single-output
SNR	signal-to-noise ratio
S/N	signal-to-noise ratio
SS	spatial spectrum
SSA	shared spectral assignment
SU	single unit (user)
TD	time delay
TDD	time-division duplexing
TDMA	time-division multiple access
TOA	time of arrival
UE	user element
WCDMA	wideband code-division multiple access

Part I

Objective

1

Overview of Wireless Networks – From 2G to 4G

Scanning the existing literature published during the recent two decades and related to the description of the wireless multiple access technologies, we notice that there are a lot of excellent works (see, for example, Refs. [1–22]), in which the multichannel, multiuser, and multicarrier accesses were described in detail for cellular and noncellular networks before and beyond third (3G) generation. However, all these works mostly described the corresponding techniques and technologies via a prism of additive white Gaussian noise (AWGN) and less via a prism of multiplicative noise that depend on fading phenomena, fast and slow, usually occurring in the wireless networks: terrestrial, atmospheric, and ionospheric [21, 22]. In other words, most of the excellent books had ignored the multiplicative noise caused by fading phenomena, which, as was shown in [21, 22], plays the main role in degradation of operational characteristics of any wireless network, such as grade of service (GoS), dealing with service of a lot of subscribers located in areas of service with a dense layout of users and quality of service (QoS), dealing with information data parameters sent and received by individual subscriber, such as the capacity, spectral efficiency, and bit error rate (BER) of data stream passing any wireless and wired communication link.

Thus, in [1–20], the authors dealt mostly with classical AWGN channels or channels with the interuser interference (IUI). As was shown there, the "response" of such channels is not time- or frequency varied, that is, such propagation channels were not time or/and frequency dispersive. In [21, 22] the authors described the main features of the multiplicative noise caused by slow and fast fading that occur in terrestrial, atmospheric, and ionospheric wireless communication links and networks. As was shown in [21, 22], the aspects of fading are very important for predicting the multiplicative noise in various radio channels, terrestrial, atmospheric, and ionospheric, for the purpose of increasing the efficiency of land–land, land–aircraft, and land–satellite communication networks. The proposed approaches were then extended for description of multimedia and optical communications based on stochastic, and other statistical, models [23–27] and on usage of special nonstandard matrices [28, 29].

Advanced Technologies and Wireless Networks Beyond 4G, First Edition.
Nathan Blaunstein and Yehuda Ben-Shimol.
© 2021 John Wiley & Sons, Inc. Published 2021 by John Wiley & Sons, Inc.

Thus, in land communication channels, due to multiple scattering, diffraction, and scattering or diffuse reflection, the channel becomes frequency selective. If one of the antennas of the subscriber, or of the base station, is moving, the channel becomes both a time- and frequency-dispersive channel. As a result, the radio signals traveling along different paths of varying lengths cause significant deviations in signal strength (in volts) or power (in watts) at the receiver. This interference picture is not changed with time and can be repeated in each phase of a radio communication link between the base station (BS) and the stationary subscriber. As for a dynamic channel, when either the subscriber antenna is in motion or the objects surrounding the stationary antennas move, the spatial variations of the resultant signal at the receiver can be seen as temporal variations, as the receiver moves through the multipath field (i.e. through the interference picture of the field strength). In such a dynamic multipath channel, a signal fading at the mobile receiver occurs in the time domain. This temporal fading relates to a shift in frequency radiated by the stationary transmitter. In fact, the time variations or dynamic changes of the propagation path lengths are related to the Doppler shift, denoted by $f_{d_{max}}$, which is caused by the relative movements of the stationary BS and/or the moving subscriber (MS). As was defined in [21, 22], the total bandwidth due to Doppler shift is $B_d = 2f_{d_{max}}$. In the time-varied or dynamic channel, for any real time t there is no repetition of the interference picture during the crossing of different field patterns by the MS at each discrete time of his movements. Thus, in Table 1.1, some characteristic parameters, such as introduced above, Doppler shift, and the delay spread, σ_t, are presented for the ionospheric and terrestrial radio communication links. According to these two main parameters of frequency- and time-dispersive channel "response," additionally two other parameters, B_c and T_c, the coherency bandwidth of channel and the time of coherency, respectively, are usually introduced to define the type of fading occurring in the desired communication channel: frequency-selective or flat (see definitions and relations between the parameters in [21]). As was shown in [21, 22], due to the "time dispersion" and "frequency dispersion" of each specific wireless communication channel, the signal data, as a stream of sequences of symbols (e.g. bits), can be corrupted by fading, and finally, a new phenomenon called intersymbol interference (ISI) is observed at the receiver. Moreover, in multiple accesses servicing, IUI is currently observed.

Table 1.1 Characteristics of fading parameters in different channels

Channel	σ_τ (s)	$f_{d_{max}}$ (Hz)
Ionospheric (HF)	$\sim 10^{-3}$	50–150
Atmospheric (HF/VHF)	$\sim 10^{-2}$	5–40
Land (VHF/UHF)	$\sim 10^{-6}$–10^{-5}	10–100

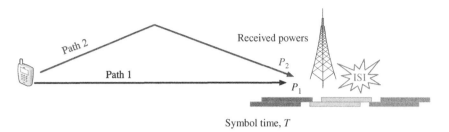

Symbol time, T

Figure 1.1 ISI caused by multipath fading phenomena

In Table 1.1, in this situation, a new "artificial noise" takes place, which causes the so-called IUI or interchannel interference (ICI). An example of how multipath fading causes ISI is shown in Figure 1.1. To overcome such kinds of effects caused by multiplicative noise, some canonical techniques were introduced in modulation schemes of current networks, defined and briefly described in [22], such as the spread spectrum modulation techniques direct sequence spread sectrum (DS-SS), frequency hopping spread spectrum (FH-SS), and time hopping spread spectrum (TH-SS)).

The work in [22] briefly explains how the IUI in the classical multiple access technologies can be overcome: CDMA (code-division multiple access) on the basis of DS-SS modulation, FDMA (frequency-division multiple access) on the basis of FH-SS modulation technique, and TDMA (time-division multiple access) on the basis of TH-SS modulation. Reference [22] also briefly introduced these techniques based on space, time, frequency, and polarization diversities for multibeam adaptive antenna applications. In this section, we introduce the orthogonal frequency-division multiplexing (OFDM) techniques and the corresponding orthogonal frequency-division multiple accesses (OFDMA), occurring in the frequency domain, as well as the orthogonal time-division multiple access (OTDMA), occurring in the time domain.

In Chapter 2, we briefly introduce the existing canonical and modern recently performed networks via their historical perspective, such as the Global System for Mobile Communications (GSM), the wireless personal area network (WPAN), also called Bluetooth, the wireless local area network (WLAN), related to the wireless fidelity (Wi-Fi) system, the wireless metropolitan area network (WirelessMAN or WiMAX), and the long-term evolution (LTE) standards. All these systems and technologies cover the time period of the past four decades in wireless generation's developments – from the past second generation to the new fourth generation. It is important to notice that all modulation techniques, the conventional CDMA/TDMA/FDMA and advanced OFDM/OFDMA/OTDMA, related to the above networks, fully depend on the fading phenomena that occur in such networks.

We do not focus in the description of the respective current and advanced protocols, such as 802.15, 802.11, and 802.16, for LTE releases, which are usually used in the above networks, as well as on the architecture of these networks, because these aspects are beyond the scope of this book and are fully described in other Refs. [1–9, 11–20]. At the same time, based on the fading parameters introduced above, we show the advantages and disadvantages of the corresponding techniques and propose for practical applications more attractive and advanced technologies.

In other words, Chapter 2 illuminates the current wireless networks and the corresponding technologies before 4G and 5G in their brief overview. In Chapter 3, we introduce some advanced diversity techniques adapted for the multicarrier accessing networks. Chapter 4 describes the advanced multiple-input-multiple-output (MIMO) spatial-time diversity and spatial multiplexing techniques, focusing the special attention on how fading phenomena affect the capacity and spectral efficiency of MIMO channels. Fading propagation effects are described in terms of the unified stochastic approach introduced in [21, 22] for land communication networks. In Chapter 5, we introduce the femtocell–microcell and femtocell–macrocell (indoor/outdoor) configurations for different types of femtocell advanced deployment strategies. Chapter 6 shows advances of the combined femtocell–microcell layout with modern concept of MIMO/LTE for future performance in 4G and 5G technologies. In previous books [22, 23], based on a general stochastic model, we describe in detail the present operative parameters of the multibeam adaptive antennas in the angle-of-arrival (AOA), time-of-arrival (TOA) or time-delay, and frequency (Doppler) domains. Finally, in Chapter 7, based on the general stochastic multiparametric approach, briefly described in Chapter 4, the main technique – how to localize, from the signal's distribution in the time-delay and Doppler spread domains, the exact position of any subscriber located in multiuser land-atmospheric communication link – is presented.

References

1 Jakes, W.C. (1974). *Microwave Mobile Communications*. New York: Wiley.
2 Lee, S.C.Y. (1989). *Mobile Cellular Telecommunication Systems*. Hoboken, NJ: McGraw-Hill.
3 Steele, R. (1992). *Mobile Radio Communication*. IEEE Press.
4 Proakis, J.G. (1995). *Digital Communications*, 3e. New York: McGraw-Hill.
5 Stuber, G.L. (1996). *Principles of Mobile Communications*. Boston, MA: Kluwer Academic Publishers.
6 Peterson, R.L., Ziemer, R.E., and Borth, D.E. (1995). *Introduction to Spread Spectrum Communications*. Hoboken, NJ: Prentice Hall PTR.
7 Rappaport, T.S. (1996). *Wireless Communications: Principles and Practice*, 2e in 2001. Hoboken, NJ: Prentice Hall PTR.

8 Steele, R. and Hanzo, L. (1999). *Mobile Communications*, 2e. Chichester: Wiley.

9 Li, J.S. and Miller, L.E. (1998). *CDMA Systems Engineering Handbook.* Boston, MA and London: Artech House.

10 Saunders, S.R. (2001). *Antennas and Propagation for Wireless Communication Systems.* Chichester: Wiley.

11 Burr, A. (2001). *Modulation and Coding for Wireless Communications.* Hoboken, NJ: Prentice Hall PTR.

12 Molisch, A.F. (ed.) (2000). *Wideband Wireless Digital Communications.* Chichester, England: Prentice Hall PTR.

13 Paetzold, M. (2002). *Mobile Fading Channels: Modeling, Analysis, and Simulation.* Chichester: Wiley.

14 Simon, M.K., Omura, J.K., Scholtz, R.A., and Levitt, B.K. (1994). *Spread Spectrum Communications Handbook.* New York: McGraw-Hill.

15 Glisic, S. and Vucetic, B. (1997). *Spread Spectrum CDMA Systems for Wireless Communications.* Boston, MA and London: Artech House.

16 Dixon, R.C. (1994). *Spread Spectrum Systems with Commercial Applications.* Chichester: Wiley.

17 Viterbi, A.J. (1995). *CDMA: Principles of Spread Spectrum Communication, Addison-Wesley Wireless Communications Series.* Reading, MA: Addison-Wesley.

18 Goodman, D.J. (1997). *Wireless Personal Communication System.* Reading, MA: Addison-Wesley.

19 Schiller, J. (2003). *Mobile Communications, Addison-Wesley Wireless Communications Series*, 2e. Reading, MA: Addison-Wesley.

20 Molisch, A.F. (2007). *Wireless Communications.* Chichester: Wiley.

21 Blaunstein, N. and Christodoulou, C. (2007). *Radio Propagation and Adaptive Antennas for Wireless Communication Links*, 1e. Hoboken, NJ: Wiley.

22 Blaunstein, N. and Christodoulou, C. (2014). *Radio Propagation and Adaptive Antennas for Wireless Communication Networks – Terrestrial, Atmospheric and Ionospheric*, 2e. Hoboken, NJ: Wiley.

23 Hadar, O., Bronfman, I., and Blaunstein, N. (2017). Optimization of error concealment based on analysis of fading types, Part 1: Statistical description and error concealment of video signals. *J. Inf. Control Syst.* 86 (1): 72–82.

24 Hadar, O., Bronfman, I., and Blaunstein, N. (2017). Optimization of error concealment based on analysis of fading types, Part 2: Modified and new models of video signal error concealment. Practical simulations and their results. *J. Inf. Control Syst.* 86 (2): 67–76.

25 Sun, M.T. and Reibman, A.R. (2001). *Compressed Video over Networks.* New York: Marcel Dekker.

26 Doshkov, D., Ndjiki-Nya, P., Lakshman, H. et al. (2010). Towards efficient intra prediction based on image inpainting methods. *28th Picture Coding Symposium*, PCS2010, Nagoya, Japan (8–10 December 2001), p. 6.

27 Chen, B.N. and Lin, Y. (2006). Selective motion field interpolation for temporal error concealment. *International Conferencde on Computer and Communication Engineering 2006* (ICCCE 2006), Kuala Lumpur, Malaysia (9–11 May 2006).

28 Hadar, O., Huber, M., Huber, R., and Greenberg, S. (2005). New hybrid error concealment for digital compressed video. *EURASIP J. Appl. Signal Process.* 12: 1821–1833.

29 Blaunstein, N., Arnon, S., Zilberman, A., and Kopeika, N. (2010). *Applied Aspects of Optical Communication and LIDAR*. Taylor and Francis Group.

2

Terrestrial Wireless Networks Based on Standard 2G and 3G Technologies

As we mentioned, at the beginning of the book, we do not get into the details of protocols and architectures of the various current (3G) and modern (4G) wireless networks. Instead, we refer the reader to [1–22], where the corresponding networks, techniques, and protocols are fully described. At the same time, we note that the most modern wireless networks and the corresponding protocols are based on the current and advanced techniques of signal processing (e.g. modulation techniques).

2.1 Bluetooth-WPAN Networks

The wireless personal area network (WPAN), also known as Bluetooth (BT), was originally created by the Ericsson company (Sweden) in 1998 before other companies started to launch this system and the corresponding technology and protocols. The WPAN system was named "Bluetooth" according to the name of the Danish king Harald Bluetooth, living at the tenth century. BT technology is based on the 802.15 protocol and is called the 802.15.1 protocol. BT was designed for management and control of low-cost, low-power radio devices, operating within small local areas (up to tens of meters), which allows stable communication at short distances between personal devices such as notebooks, cellular phones, and personal computers. Such type of small-range area network was referred to as "piconet" in [5].

Currently, the BT technology allows sending a data stream through each channel with a maximal rate of 1 Mbps, that is, it allocates for each channel of WPAN system a nominal bandwidth of 1 MHz. The WPAN system operates at a carrier frequency of 2.4 GHz, using the frequency-hopping spread spectrum (FH-SS) modulation technique, described in detail in [4]. Thus, the whole bandwidth, consisting of 79 hopping channels, is ranged from 2.402 to 2.480 GHz with minimum hopping range of six channels. The frequency hopping is performed 1600 times per second. Piconet is presented as a cluster of up to eight radio devices, one with the role of a "master," and the rest are "slaves." The WPAN-BT system is based on TDD (time-division duplexing)

Advanced Technologies and Wireless Networks Beyond 4G, First Edition.
Nathan Blaunstein and Yehuda Ben-Shimol.
© 2021 John Wiley & Sons, Inc. Published 2021 by John Wiley & Sons, Inc.

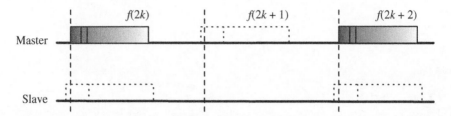

Figure 2.1 Stream of information packets via a Bluetooth network

technique, according to which the channel is divided into timeslots of 625 µs each. Such a division in the time domain allows the transmission of a data packet to be transmitted in a single timeslot. The master sends packets to slaves only in even-numbered timeslots, while a slave sends packets to the master only in odd-numbered slots, as shown in Figure 2.1.

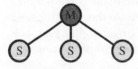

Figure 2.2 Distribution of devices in one piconet: one master and from three to seven slaves

The configuration for each piconet is as follows: one device is designated as the master, and up to seven devices are designated as the slaves (see Figure 2.2). In such a spatial configuration in local area of service, two different piconets are connected via bridges where a device can be the "master" in one piconet and a "slave" in another piconet (in Figure 2.3, we present example of two overlapping piconets).

Finally, seven slaves can be activated simultaneously in one piconet. If there are more than seven slaves in one piconet, then the rest of the slaves must be parked. In such configuration, the maximum number of parking slaves cannot exceed 255 per piconet (as shown in Figure 2.4).

Let us now briefly state the advantages and disadvantages of the WPAN-BT system.

The advantages are:

- An effective and inexpensive wireless solution for both data and voice at short distances from the receiver.
- Applicable for stationary and mobile environments.
- No setup is required.

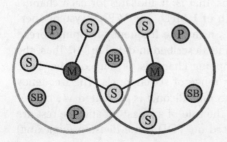

Figure 2.3 Distribution of masters and slaves between two piconets

Figure 2.4 Final configuration of several piconets covering an area of service

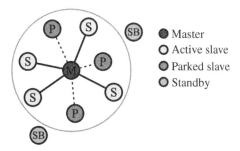

- Low power consumption.

The disadvantages are:

- Short antenna range (up to 10–20 m).
- Low data rate.
- Possible interference from other networks operating at the same frequency bands.
- No data security.

2.2 Wi-Fi–WLAN Networks

In 1990, the Institute of Electrical and Electronics Engineers (IEEE) established the 802.11 working group to create a wireless local area network (WLAN) providing a set of standards for WLANs. The wing ".11" refers to a subset of the 802 group which is the WLAN working group [6, 7]. Then, WLANs were associated with Wi-Fi (Wireless Fidelity) networks. The IEEE 802.11 working group and the Wi-Fi Alliance [23] came out as the key groups in creating different 802.11 protocol standards. Thus, we also associate WLAN technologies and the corresponding protocols with Wi-Fi networks.

WLAN systems, which were designed for servicing pico/micro cells (i.e. up to 1–2 km), are based on the standard protocol 802.11. Its physical layer is based on the standard FH-SS modulation technique (see [3]). Modern WLANs are now widely accepted and used in private and local commercial areas to support subscribers, stationary and mobile, with special terminals, called access points (APs).

Examples of how mobile subscribers connect to an AP via Wi-Fi/WLAN network are shown schematically in Figure 2.5, where, for WLAN networks, a medium access control (MAC) technique is used for providing quality-of-service (QoS) in packet-switched services of multiple subscribers located in picocell and microcell local areas (the corresponding protocol is called IEEE 802.11 MAC standards) [8–11]. The main goal of IEEE 802.11 MAC was to support QoS for real-time services, such as telephony, multimedia (video and audio) communications.

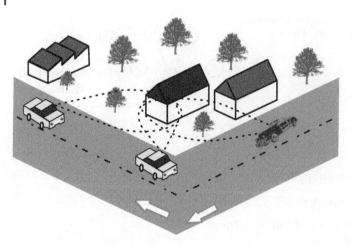

Figure 2.5 Road-Wi-Fi network

One such service is Voice over Internet Protocol (VoIP) – a popular network application, where the network converts voice data to digital form and vice versa. However, since today's VoIP calls are possible in a WLAN environment, there are a number of factors that negatively affect the use and acceptance of VoIP. Current WLANs have limited ability to support multimedia communications. Therefore, as will be mentioned briefly below, QoS provisions must be incorporated with the current WLAN systems to support the requirements of real-time services that are required for applications such as VoIP. For further reading on advanced techniques to supporting real-time voice services in WLAN systems, the reader is referred to Refs. [12–19]. Below, we briefly present overview results of these research works. We note that different modifications of IEEE 802.11 technology and its protocol were introduced during the past two decades, such as protocols 802.11a/b/g and 802.11n/e, on the basis of orthogonal frequency-division multiplexing/orthogonal frequency-division multiple accesses (OFDM/OFDMA) modulation techniques.

Why such a broad set of 802.11 standards were created? To answer this question, let us briefly describe some of the popular standards of 802.11 technologies. This will help the reader to understand how the standard technologies, networks, and the corresponding protocols allow the designers of WLAN/Wi-Fi systems to increase the efficiency of grade of service (GoS) and QoS of WLANs and how to eliminate the intersymbol interference (ISI) and the intercarrier interference (ICI) in each channel (e.g. each carrier) of the desired system.

WLAN and the corresponding 802.11b technology and protocol were adopted in 1999 by moving from fixed wireless network to dynamic wireless network. The protocol 802.11b is based on direct sequence spread spectrum (DS-SS), which is described later in Chapter 3. 802.11b allows the connection of hundreds of computers and devices using a 2.4- GHz carrier frequency. This technology was focused on the physical layer and data link layer simultaneously [12]. 802.11b was designed for ranges of up to 100 m and supported the transmissions of data streams with a maximal rate of 11 Mbps. Fallback rates of 1, 2, and 5.5 Mbps were supported for bad channel conditions due to noise, clutter conditions, distance from APs, and so forth.

The logical structure of 802.11 technologies allows the usage of FH-SS modulation at the physical layer, in addition to DS-SS modulation, combining with logical link control (LLC) layer and MAC layer. The LLC provides the addressing and data link control, independently from any topology and medium, and connecting to MAC, which provides the access to wireless medium in the presence of multiple users.

Using 14 nonoverlapping channels, each of 22 MHz bandwidth, placed 5 MHz apart (e.g. channel 1 is placed at central frequency of 2.412 GHz, channel 2 is at 2.417 GHz, and so on, up to channel 14, which is placed at 2.477 GHz). Such logical structure of 802.11 protocols results with multiple benefits such as:

- Wide coverage range in an indoor/outdoor picocell/microcell environment.
- Free and stable work, both with stationary and mobile subscribers.
- Possibility to work with other picocell networks, such as WPAN, using the same 2.4 GHz frequency band (see below).
- Scalability and security for each subscriber located in the area of service.

2.2.1 Integrated WLAN and WPAN Networks

At the same time, since the RF band of WLANs, based on the 802.11b protocol, is the same as in other networking technologies, such as BT (WPAN), 2.4 GHz cordless phones, interference between them is usually observed, decreasing the efficiency of these networks. Furthermore, low QoS provisions for multimedia content and a lack of interoperability with voice sensors and devices decrease the efficiency of 802.11b technology. This generated the idea of combining the advantages of WPAN (e.g. BT) and those of WLAN based on the 802.11b standard protocol, because they both operate at the same 2.4 GHz frequency band and can use the same modulation technique [13]. Moreover, because BT is quick, operates at short ranges, and protects data and voice transfer, it can obey the limitations existing in WLANs, such as limitations for voice and multimedia data. At the same time, the 802.11b technology was designed for infrequent mobility IP-based data transmission with high data rate at medium ranges of service (up to 100 m).

As for the interference between BT and WLAN 802.11b technologies, this problem was resolved at both the physical and MAC layers of the 802.11b network. Finally, combining the small area of operation, the fast FH-SS (1600 hopes per second) to avoid interference by jumping to another frequency and small packet size of BT, compared with the large WLAN packets, an increase of throughput, QoS, and GoS of the combined system was finally obtained [24].

During the past decade, a new protocol – IEEE 802.11n – was introduced for supporting modern Wi-Fi systems, as well as for multibeam adaptive antennas implementation (called the multiple-input-multiple-output [MIMO] system), which is based on the combined OFDM and MIMO technologies described in Chapters 3 and 4. These modern systems operate at the 2.4–2.6 GHz frequency band and cover microcells of several hundred meters with a maximum data rate (e.g. capacity) of up to 100 Mbps [25]. In most of the advanced technologies, each subscriber is associated with an AP, and the corresponding multibeam BS antenna is involved in the multiple access servicing via the so-called the infrastructure mode [13]. A final version of 802.11n standard was published in 2009. The corresponding Wi-Fi network based on this technology spreads the frequency band from 2.4 to 3.5–5 GHz with bandwidth 20–40 MHz. The MIMO system for this standard consists of 4×4 antenna elements.

2.2.2 Enhancement of the WLAN Technology

Recently, the enhancement of the existing protocols, called IEEE 802.11e standard, was performed for VoIP [15–21]. The VoIP technology has advanced rapidly during recent years. The problem was to make it suitable for WLAN technology. The typical VoIP WLAN architecture is presented in Figure 2.6.

The voice conversation happens through the AP. The analog voice signal, after compression, encoding, and packetization procedures at the WLAN layers, is transmitted over the networks and finally to the receiver end terminal. At the receiver, the effects of signal delay and delay jitter are compensated by a special buffer. Finally, the receiver depacketizes and decodes the received packets to recover the original voice signal. As was mentioned in [18], "one major challenge for VoIP WLAN is the QoS provisioning." As was mentioned above, WLANs were designed for high-rate data traffic, and they may experience bandwidth inefficiency, dealing with low-rate and delay-sensitive voice traffic.

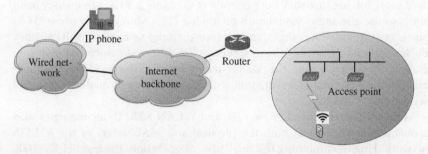

Figure 2.6 VoIP WLAN standard architecture

Therefore, the idea is to enhance the QoS support capability of current WLAN standard protocols, from 802.11a to 802.11n, by a new more popular standard protocol 802.11e [18, 19]. The 802.11n enhanced technology is superior to the previous ones, since it allows transmitting low-power signal data with shorter time delay and supports better duplex transmission for applications such as VoIP, by sending upstream and downstream voice signals separately, via network channels [9, 11, 19, 22].

Using beam-forming antennas, as additional attributes of this technology, the 802.11n network allows eliminating multipath fading effects and, therefore, eliminates ISI and ICI that usually occur in wireless communication in environments with fading (see details in [23–84]).

2.3 WiMAX Networks and 802.16 Technologies

The goals of the IEEE 802.16 standard (also commercially known as WiMAX) were the design and implementation of broadband wireless systems that operate on the basis of adaptive multibeam or phased array antennas for macrocell servicing (up to tens of kilometers) [85–92]. A WiMAX antenna can cover metropolitan areas of several tens of kilometers for fixed stations and up to 10 km for mobile stations. Therefore, initially (on April 2002) the IEEE standard 802.16-2001 was defined as wireless metropolitan area network (WMAN) [86].

A WMAN offers an alternative to wire-line communication networks which use cables or fiber optics with their modems and digital subscriber line (DSL) links. WiMAX networks have a huge capacity to address broad geographic areas without additional infrastructure required in cable links installation in each individual site or for each individual subscriber, as shown in Figure 2.7. In such scenario, WiMAX technology brings the network to subscribers located inside which are connected with conventional indoor networks such as Ethernet or wireless local area networks (LANs).

The fundamental design of MAC standard allows WiMAX systems to serve all individual users, moving along roads or at home, as shown in Figure 2.8. This can be done despite the fact that they use quite different physical layers and are located in different environmental conditions.

With MAC technology expanding in this direction, it is important to emphasize that the 802.16 MAC standard technology could accommodate all connections with full QoS and increase of GoS.

The signal processing technique implemented in a WiMAX system is based on OFDM/OFDMA modulation techniques, which will be described later in Section 2.2, and operates at frequencies from 2 to 10 GHz. In Figure 2.9 we illustrate the OFDMA technique, converting its presentation into a view of "user applications." Thus, according to the illustration, each terminal occupies a subset of subcarriers. The subset is called an OFDMA traffic channel, and each traffic channel is assigned exclusively to one user at any time.

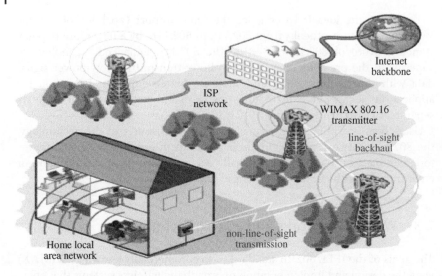

Figure 2.7 WMAN – metropolitan area network (extracted from Internet)

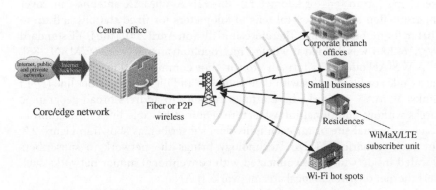

Figure 2.8 Full configuration of WiMAX network using MAC 802.16 standard

A WiMAX antenna can transfer information data with a maximum rate of up to 70 Mbps. The main goal of such technology is to handle any effects of NLOS in urban and suburban environments that usually occur in the built-up terrestrial environments (see [3]). The main features of WiMAX networks are [85–90]:

(a) Using advanced OFDMA technique;
(b) The bandwidth varies from 1.25 to 28 MHz;
  - Using TDD and FDD techniques (see definitions in [3]);
  - Using MIMO antenna systems based on a beam-forming technology of each element of BS, AP, and MU antenna performance.
(c) Using advanced signal modulation techniques;

Figure 2.9 The OFDMA concept

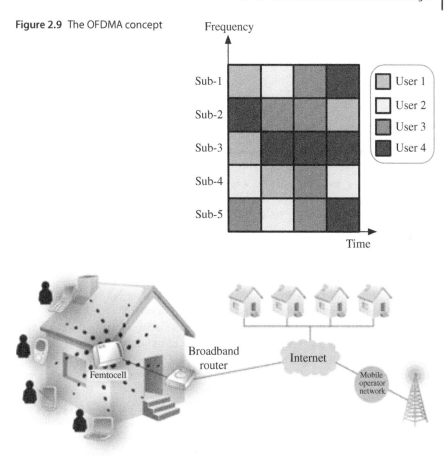

Figure 2.10 Schematic presentation of Femtocell-Wi-Fi–WiMAX configuration

(d) Using advanced coding techniques such as space–time coding and turbo coding.

Recently, the WiMAX technology was deployed to operate simultaneously with macrocell BS antennas and Femto-AP antennas (see Figure 2.10). We describe the possibility of integrating these configurations in Chapter 4.

2.3.1 Integrated Wi-Fi–WiMAX Networks

A tendency of integration of narrow-range Wi-Fi networks with a wide-range WiMAX network, operating at different rates and having different mobility support (see Figure 2.11), is sensitive to blocking of users calls, dynamic spectrum assignment for each user, stationary or mobile, and to energy-efficient handover schemes with the geographic mobility awareness (HGMA).

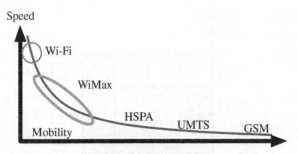

Figure 2.11 Comparison between operational characteristics, speed and mobility, of WiMAX and Wi-Fi systems

In Figure 2.11 we also introduce a GSM network that supports the high mobility of any subscriber, but having low bit/packet rate within the desired user's channel. Conversely, Wi-Fi and WiMAX technologies allow obtaining high-rate data streams, but they are not adapted for communication under high-mobility conditions. Systems that are based on high-speed packet access (HSPA) or universal mobile telephony systems (UMTS) are also limited in terms of data stream transfer via user's channels with high speed.

The problems of integration of different systems having limited possibilities either in mobility or in speed, such as Wi-Fi and WiMAX systems, were discussed in [93–98], where the main goal of the researchers was to decrease the blocking probability and increase the efficiency of handover schemes and frequency spectrum sharing among user's channels. Following [93–98], we briefly introduce the reader to some of the problems and the algorithms used to overcome them.

A Wi-Fi/WiMAX integrated network was proposed to achieve high-quality communication using Wi-Fi and WiMAX as complementary access resources. The integrated network, according to the researchers' main aim, will enable to support a load balancing between Wi-Fi and WiMAX using each system selectively in response to the demands of subscribers, stationary and/or mobile, and the usage status of each system. According to such an idea, in the integrated Wi-Fi/WiMAX network each wireless system will use the spectrum band prescribed by law, so that even if the WiMAX system has unused spectrum temporarily, it cannot be used by Wi-Fi wireless systems.

The first problem, investigated in [96], was to find an effective spectrum sharing method for Wi-Fi/WiMAX integrated mesh networks. Resolving the problem of spectrum sharing in Wi-Fi networks allows connecting the Wi-Fi mesh network to a WiMAX base station (BS), obtaining an increase of throughput. The problem is that in a Wi-Fi mesh network, several Wi-Fi APs are interconnected by wireless links and the communication with the backbone network transits through the gateway AP that is connected by wire cables. It was expected to reduce the cost of infrastructure and to adapt it, not only to urban but also to rural areas. In addition, in a multichannel, multiinterface mesh network, where each AP can use two or more channels simultaneously, it was found possible to increase the network capacity by dynamic

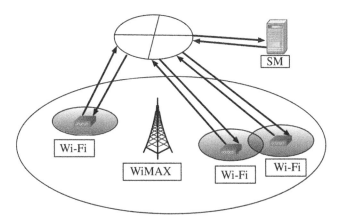

Figure 2.12 Configuration where Wi-Fi APs are interconnected by backbone (denoted by the circle above the WiMAX BS). The backbone is connected to the local Wi-Fi APs by wired technologies

channel assignment for each wireless link. However, when many mobile users communicate with the backbone network, the network throughput decreases due to congestion around gateway APs [96]. To overcome this problem, a Wi-Fi/WiMAX integrated mesh network was proposed in [96], where WiMAX is used as backhaul for the Wi-Fi mesh network. In such a combined network, there are two kinds of gateway APs: one is a traditional gateway AP, directly connected to the backbone by a wired cable, and the other is an AP, which is wirelessly connected with the WiMAX BS and functions as a gateway.

All transfers are under the control of a special server: the "spectrum manager" (SM), as shown in Figure 2.12. Such a method allows improving the throughput of the combined system and increases the number of gateways for wireless communication between Wi-Fi APs and the WiMAX BS, as well as of gateway APs without laying new wire cables.

Also, dynamic spectrum assignment, based on call blocking probability prediction in an integrated Wi-Fi/WiMAX network, was investigated in [97]. The main idea of this research was to allow Wi-Fi systems to temporally use a spectral band of WiMAX systems, in an integrated network of operation stages. Thus, because the Wi-Fi system uses the spectrum in units of 20 MHz, the WiMAX system divides its spectrum into channels of 20 MHz and assigns one of them to the Wi-Fi APs, which leads to more effective use of spectrum in the integrated network. To achieve this, a channel in the WiMAX system needs to be assigned to as many Wi-Fi systems as possible, as long as they are not adjacent, as shown in Figure 2.13. Specifically, a WiMAX system provides 74.8 Mbps per channel, and a Wi-Fi system provides 54 Mbps per channel at maximum.

Therefore, if two or more Wi-Fi APs use one channel of the WiMAX system, the spectrum utilization efficiency can be enhanced for the whole integrated

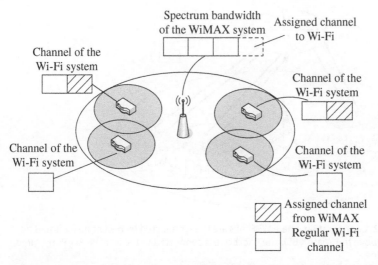

Figure 2.13 Another configuration of Wi-Fi APs incorporated with backbone (denoted by dark circles around each WiMAX BS). Each backbone can operate both with WiMAX and Wi-Fi channel

network. The proposed method is based on the predicted numbers of blocked calls, an analytical analysis of which is fully presented in [98]. The proposed method [96–98] introduces an effective dynamic spectrum assignment, where the same spectrum can be repeatedly used by assigning a channel of the WiMAX system to two or more Wi-Fi systems without causing interference between adjacent Wi-Fi APs.

At the same time, to handle any handoff problems, in [98] the HGMA was proposed by considering the past handover patterns of mobile devices (MDs) and by conservation of the energy of handover devices.

2.4 LTE Current Technologies

The long-term evolution (LTE) technology, networks, and the corresponding protocols were developed recently to increase the capacity and speed of wireless data passing wireless communication links and networks, using modern hardware (compared to WiMAX networks) and advanced digital signal processing techniques [99–111]. LTE was defined in 2009 by the 3rd Generation Partnership Project (3GPP) as a highly flexible broadband radio system with high user data rate (up to 30 Mbps), with a data stream and radio sensors/networks delay not exceeding 5 ms, with simple network architecture, efficient spectra allocations, and so on. As was mentioned in [105]: "LTE is designed to meet carrier needs for high-speed data and media transport as well as high-capacity voice support well into the next decade."

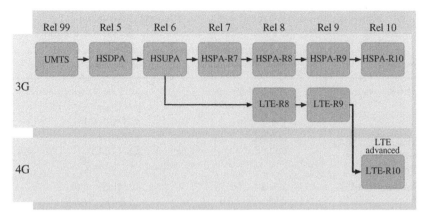

Figure 2.14 The evolution of LTE releases (Rels) during recent years

Moreover, the main goal of the first nine releases of LTE was to support both frequency-division duplexing (FDD) and TDD combined with a wideband system in order to achieve a large number of various spectra allocations [104, 105]. All these nine standard releases (see Figure 2.14) were implemented recently to overcome the well-known WiMAX technologies, such as enhanced IEEE 802.16e, performed in 2005. One can see from comparing the two systems (see Table 2.1, extracted from Ethernet) that, for example, LTE E-UTRAN has much better parameters regarding data speed and data protection within communication channels.

In Figure 2.15, replicated from [105], an example of the arrangement of the 10 subframes into a one radio frame is illustrated.

The LTE supports FDD and TDD simultaneously (see Figure 2.16, extracted from [105]) to send to each user the information data, as well as the control signals, with small delay between the short subframe duration of 1 ms. Recently, LTE was introduced to support the systems that can be considered as a continuous evolution from earlier 3GPP networks, such as TD-SCDMA (time-division synchronous code-division multiple access) and wide-band code-division multiple access (WCDMA) combined with HSPA. The current Releases 8 and 9 of LTE technology, denoted sometimes as 3GPP-LTE (or E-UTRAN), include many of the features of 4G systems. Therefore, they were considered as the best candidates for 4G generation of networks and as a major step toward the advanced international mobile telephony (ITM-Advanced) [101–119], namely, LTE Release 8 was performed for single-user (SU) network for servicing each user equipment (UE). Its arrangement is presented schematically in Figure 2.17.

It has the following characteristics [118]:

- Combining TDD and FDD modes and OFDM technique in downlink (DL) and SC (single carrier), FDMA technique in the uplink (UL), using adaptive

Table 2.1 Comparison between enhanced WiMAX 802.16e and LTE technologies

Attribute	Mobile WiMAX (IEEE 802.16e-2005)	3GPP LTE(E-UTRAN)
Core network	WiMAX forum All-IP network	UTRAN moving toward All-IP evolved UTRA CN with IMS
Access technology Downlink (DL) Uplink (UL)	OFDMA OFDMA	OFDMA SC-FDMA
Frequency band	2.3–2.4 GHz 2.49–2.69 GHz 3.3–3.8 GHz	Existing and new frequency bands (~2 GHz)
Bit-rate/site DL UL	75 Mbps (MIMO 2Tx × 2Rx) 25 Mbps	100 Mbps (MIMO 2Tx × 2Rx) 50 Mbps
Channel bandwidth (BW)	5, 8.75, 10, 20 MHz	1.4–20 MHz
MIMO DL UL	2Tx × 2Rx 1Tx × NRx (collaborative SM)	2Tx × 2Rx 2Tx × 2Rx
Cell capacity	100–200 users	> 200 users @ 5 MHz > 400 users for larger BW
Spectral efficiency	3.5 (bps/Hz)	5 (bps/Hz)

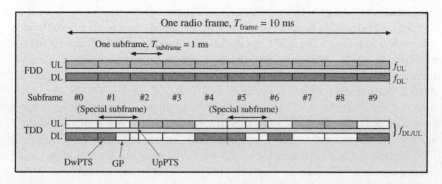

Figure 2.15 LTE frame structure where the total frame is divided at 1-ms subframes, both for the uplink (UL) and downlink (DL) system configurations (replicated from [105])

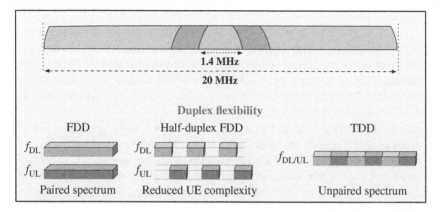

Figure 2.16 LTE frame structure where the total frame is divided at 1-ms subframes, both for the uplink (UL) and downlink (DL) system configurations (replicated from [103])

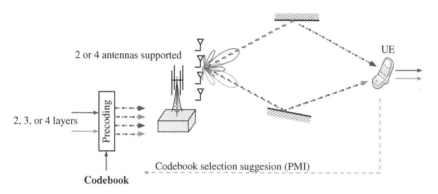

Figure 2.17 LTE frame structure where the total frame is divided at 1-ms subframes, both for the uplink (UL) and downlink (DL) system configurations (replicated from [105, 118])

modulation and coding such as QPSK/16QAM/64QAM in both DL and UL channels.

- Data rate for 20-MHz bandwidth: 100 Mbps in DL and 50 Mbps in UL channel. Spectral efficiency for 20-MHz bandwidth: 5(bits/s/Hz) in DL and 2.5 (bits/s/Hz) in the UL.
- Latency (e.g. delay of data for each desired user) is less than 5 ms for small IP packets.

The LTE Release 9 standard was performed as an enhanced version of LTE Release 8 standard [106, 107], where for demodulation purposes a virtual antenna with precoded UE-specific reference signals was added. Here also, both paired and unpaired bands of the radio spectra were proposed, depending on the types of environment, rural, suburban, urban, on the built-up terrain features, and on the configuration of the bandwidth allocated for users servicing. The paired frequency bands correspond to configurations where UL and DL transmissions are assigned separate frequency bands, whereas the unpaired frequency bands correspond to configurations where UL and DL must share the same frequency band. As illustrated in Figure 2.16, the LTE technology allows for an overall system bandwidth ranging from as small as 1.4 up to 20 MHz, where the latter is required to provide the highest data rate within LTE-system communication channels.

All user terminals support the widest bandwidth. Unlike previous cellular systems of third generation mentioned above, the LTE system provides the possibility for different UL and DL bandwidths, enabling asymmetric spectrum utilization. Usage of effective and flexible spectra sharing, not only in different frequency bands but also different bandwidths, combining with efficient migration of other radio-access technologies to LTE technology, is the main key of the LTE radio access that provides a good foundation for further fourth generation evolution. In Chapter 6, we present combination of LTE advanced technology with a MIMO system that is planned to be useful for fourth and fifth generations.

References

1 Schiller, J. (2003). *Mobile Communications, Addison-Wesley Wireless Communications Series*, 2e. Reading, MA: Addison-Wesley.
2 Molisch, A.F. (2007). *Wireless Communications*. Chichester: Wiley.
3 Blaunstein, N. and Christodoulou, C. (2014). *Radio Propagation and Adaptive Antennas for Wireless Communication Networks – Terrestrial, Atmospheric and Ionospheric*, 2e. Hoboken, NJ: Wiley.
4 Krouk, E. and Semenov, S. (eds.) (2011). *Modulation and Coding Techniques in Wireless Communications*. Chichester: Wiley.
5 Specification of the Bluetooth System. http://www.bluetooth.com. (accessed 1 December 1999).
6 Junaid, M., Mufti, M., and Ilyas, M.U. (2006). Vulnerabilities of IEEE 802.11i wireless LAN. *Trans. Eng. Comput. Technol.* 11: 1305–5313.
7 IEEE 802.11 working group. http://grouper.ieee.org/groups/802/11/index.html (accessed 7 May 2020).
8 Wireless Ethernet Compatibility Alliance. http://www.wirelessethernet.org/index.html (accessed 7 May 2020).
9 Sharon, O. and Altman, E. (2001). An efficient polling MAC for wireless LANs. *IEEE/ACM Trans. Netw.* 9 (4): 439–451.
10 IEEE std. 802.11-1999 (1999). Wireless LAN Medium Access Control (MAC) and Physical Layer (PHL) Specifications.
11 Qainkhani, I.A. and Hossain, E. (2009). A novel QoS-aware MAC protocol for voice services over IEEE 802.11-based WLANs. *J. Wireless Commun. Mob. Comput.* 9: 71–84.

12 IEEE 802.11 Working Group (1999). Part 11: wireless LAN medium access control (MAC) and physical layer (PHY) specifications: higher-speed physical layer extension in the 2.4 GHz band. ANSI/IEEE Standard 802.11.

13 Zyren, J. (1999). Reliability of IEEE 802.11 High Rate DSSS WLANs in a High Density Bluetooth Environment; 802.11 Section, 8-6.

14 Perahia, E. (2008). IEEE 802.11n development: history, process, and technology. *IEEE Commun. Mag.* 46: 46–55.

15 Ni, Q., Romshani, L., and Turletti, T. (2004). A survey of QoS enhancements for IEEE 802.11 wireless LAN. *J. Wireless Commun. Mob. Comput.* 4 (5): 547–566.

16 Wang, W., Liew, S.C., and Li, V.O.K. (2005). Solutions to performance problems in VoIP over 802.11 Wireless LAN. *IEEE Trans. Veh. Technol.* 54 (1): 366–384.

17 Robinson, J.W. and Randhawa, T.S. (2004). Saturation throughput analysis of IEEE 802.11e enhanced distributed coordination function. *IEEE J. Sel. Areas Commun.* 22 (5): 917–928.

18 Wang, P., Jiang, H., and Zhuang, W. (2006). 802.11e enhancement for voice service. *IEEE Wireless Commun.* 13 (1): 30–35.

19 Perez-Costa, X. and Camps-Mur, D. (2010). IEEE 802.11e QoS and power saving features overview and analysis of combined performances. *IEEE Wireless Commun.* 17 (2): 88–96.

20 Kopsel, A. and Wolisz, A. (2001). Voice transmission in an IEEE 802.11 WLAN based access network. *Proceedings of 4th ACM International Workshop on Wireless Mobile Multimedia* (WoWMoM). Rome, Italy, pp. 24–33.

21 Veeraraghavan, M., Chocker, N., and Moors, T. (2001). Support of voice services in IEEE 802.11 wireless LANs. *Proceedings Of IEEE INFOCOM'01*, Anchorage, Alaska, Volume 1, pp. 488–497.

22 Kim, Y.-J. and Suh, Y.-J. (2004). Adaptive polling MAC schemes for IEEE 802.11 wireless LANs supporting voice-over-IP (VoIP) services. *J. Wireless Commun. Mob. Comput.* 4: 903–916.

23 Oyman, O., Nabar, R.U., Boleskei, H., and Paulraj, A.J. (2003). Characterizing the statistical properties of mutual information in MIMO channels. *IEEE Trans. Signal Process.* 51: 2784–2795.

24 Paulraj, A.J. and Kailath, T. (1994). Increasing capacity in wireless broadcast systems using distributed transmission/directional reception (DTDR). US Patent 5, 345, 599, 6 September 1994.

25 Foschini, G.J. (1996). Layered space-time architecture for wireless communication in a fading environment when using multiple antennas. *Bell Labs. Tech. J.* 1 (2): 41–59.

26 Andersen, J.B. (2000). Array gain and capacity of known random channels with multiple element arrays at both ends. *IEEE J. Sel. Areas Commun.* 18: 2172–2178.

27 Blaunstein, N. and Yarkoni, N. (2006). Capacity and spectral efficiency of MIMO wireless systems in multipath urban environment with fading. *Proceedings of the European Conference on Antennas and Propagation*, EuCAP-2006. Nice, France, pp. 111–115.

28 Tsalolihin, E., Bilik, I., and Blaunstein, N. (2011). MIMO capacity in space and time domain for various urban environments. *Proceedings of 5th European Conference on Antennas and Propagat*, (EuCAP). Rome, Italy (11–15 April 2011), pp. 2321–2325.

29 Molisch, A., Steinbauer, M., Toeltsch, M. et al. (2002). Capacity of MIMO systems based on measured wireless channels. *IEEE J. Sel. Areas Commun.* 20: 561–569.

30 Gesbert, D., Shafi, M., Shiu, D. et al. (2003). From theory to practice: an overview of MIMO space-time coded wireless systems. *IEEE J. Sel. Areas Commun.* 21 (3): 281–302.

31 Shiu, M.D., Foschini, G.J., and Kahm, J.M. (2000). Fading correlation and its effect on the capacity of multi-antenna systems. *IEEE Trans. Commun.* 48: 502–513.

32 Philippe, J., Schumacher, L., Pedersen, K. et al. (2002). A stochastic MIMO radio channel model with experimental validation. *IEEE J. Sel. Areas Commun.* 20(6): 1211–1226.

33 Gesbert, D., Boleskei, H., Gore, D.A., and Paulraj, A.J. (2002). Outdoor MIMO wireless channels: models and performance prediction. *IEEE Trans. Commun.* 50 (6): 1926–1934.

34 Boleskei, H., Borgmann, M., and Paulraj, A.J. (2002). On the capacity of OFDM-based spatial multiplexing systems. *IEEE Trans. Commun.* 50 (1): 225–234.

35 Boleskei, H., Borgmann, M., and Paulraj, A.J. (2003). Impact of the propagation environment on the performance of space-frequency coded MIMO-OFDM. *IEEE J. Sel. Areas Commun.* 21 (2): 427–439.

36 Chizik, D., Ling, J., Wolniansky, P.W. et al. (2003). Multiple-input-multiple-output measurements and modeling in Manhattan. *IEEE J. Sel. Areas Commun.* 23 (2): 321–331.

37 Paulraj, A.J., Gore, D.A., Nabar, R.U., and Boleskei, H. (2004). An overview of MIMO communications – a key to gigabit wireless. *Proc. IEEE* 92 (2): 198–218.

38 Forenza, A., McKay, M.R., Pandharipande, A. et al. (2007). Adaptive MIMO transmission for exploiting the capacity of spatially correlated channels. *IEEE Trans. Veh. Technol.* 56 (2): 619–630.

39 Foschini, G.J. and Gans, M.J. (1998). On limits of wireless communications in a fading environment when using multiple antennas. *Wireless Pers. Commun.* 6 (3): 311–335.

40 Proakis, J.G. (2001). *Digital Communications*, 4e. New York: McGraw-Hill.

41 Golden, G.D., Foschini, G.J., Valenzula, R.A., and Wolniansky, P.W. (1999). Direction algorithm and initial laboratory results using the V-BLAST space–time communication architecture. *Electron. Lett.* 35(1): 14–15.

42 Nabar, R.U., Bolcskei, H., Erceg, V. et al. (2002). Performance of multi-antenna signaling techniques in the presence of polarization diversity. *IEEE Trans. Signal Process.* 50 (10): 2553–2562.

43 Zheng, L. and Tse, D. (2003). Diversity and multiplexing: a fundamental tradeoff in multiple antenna channels. *IEEE Trans. Inf. Theory* 49 (5): 1073–1096.

44 Varadarajan, B. and Barry, J.R. (2002). The rate-diversity trade-off for linear space-time codes. *Proceedings of IEEE Vehicular Technology Conference*, Volume 1, pp. 67–71.

45 Godovarti, M. and Nero, A.O. (2002). Diversity and degrees of freedom in wireless communications. *Proc. ICASSP* 3: 2861–2864.

46 Raleigh, G.G. and Cioffi, J.M. (1998). Spatio-temporal coding for wireless communication. *IEEE Trans. Commun.* 46 (3): 357–366.

47 Wittniben, A. (1991). Base station modulation diversity for digital simulcast. *Proceedings of IEEE Vehicular Technology Conference*, pp. 848–853.

48 Seshadri, N. and Winters, J.H. (1994). Two signaling schemes for improving the error performance of frequency-division-duplex (FDD) transmission systems using transmitter antenna diversity. *Int. J. Wireless Inf. Netw.* 1 (1): 49–60.

49 Alamouti, S.M. (1998). A simple transmit diversity technique for wireless communications. *IEEE J. Sel. Areas Commun.* 16 (8): 1451–1458.

50 Tarokh, V., Seshandri, N., and Calderbank, A.R. (1999). Space-time codes for high data rate wireless communication: performance criterion and code construction. *IEEE Trans. Inf. Theory* 45 (5): 1456–1467.

51 Ganesan, G. and Stoica, P. (2001). Space–time block codes: a maximum SNR approach. *IEEE Trans. Inf. Theory* 47 (4): 1650–1656.

52 Hassibi, B. and Hochwald, B.M. (2002). High-rate codes that are linear in space and time. *IEEE Trans. Inf. Theory* 48 (7): 1804–1824.

53 Health, R.W. Jr. and Paulraj, A.J. (2002). Linear dispersion codes for MIMO systems based on frame theory. *IEEE Trans. Signal Process.* 50 (10): 2429–2441.

54 Winters, J.H. (1998). The diversity gain of transmit diversity in wireless systems with Rayleigh fading. *IEEE Trans. Veh. Technol.* 47 (1): 119–123.

55 Bjerke, B.A. and Proakis, J.G. (1999). Multiple-antenna diversity techniques for transmission over fading channels. *Proceedings of Wireless Communications Networking Conference*, Volume 3, pp. 1038–1042.

56 Heath, R.W. Jr. and Paulraj, A.J. (2005). Switching between diversity and multiplexing in MIMO systems. *IEEE Trans. Commun.* 53 (6): 962–968.

57 Chandrasekhar, V., Andrews, J.G., and Gatherer, A. (2003). Femtocell networks: a survey. *IEEE Commun. Mag.* 46 (9): 59–67.

58 Shannon, C.E. (1948). A mathematical theory of communication. *Bell System Tech. J.* 27: 379–423 and 623–656.

59 Yeh, S.-P., Talwar, S., Lee, S.-C., and Kim, H. (2008). WiMAX femtocells: a perpective on network architecture, capacity, and coverage. *IEEE Commun. Mag.* 46 (10): 58–65.

60 Knisely, D.N., Yoshizawa, T., and Favichia, F. (2009). Standardization of femtocells in 3GPP. *IEEE Commun. Mag.* 47 (9): 68–75.

61 Knisely, D.N. and Favichia, F. (2009). Standardization of femtocells in 3GPP2. *IEEE Commun. Mag.* 47 (9): 76–82.

62 Chandrasekhar, V. and Andrews, J.G. (2009). Uplink capacity and interference avoidance for two-tier femtocell networks. *IEEE Trans. Wireless Commun.* 8 (7): 3498–3509.

63 Calin, D., Claussen, H., and Uzunalioglu, H. (2010). On femto deployment architectures and macrocell offloading benefits in joint macro-femto deployments. *IEEE Commun. Mag.* 48 (1): 26–32.

64 Kim, R.Y., Kwak, J.S., and Etemad, K. (2009). WiMAX Femtocel: requirements, challenges, and solutions. *IEEE Commun. Mag.* 47 (9): 84–91.

65 Lopez-Perez, D., Valcarce, A., de la Roche, G., and Zhang, J. (2009). OFDMA femtocells: a roadmap on interference avoidance. *IEEE Commun. Mag.* 47 (9): 41–48.

66 Chandrasekhar, V., Andrews, J.G., Muharemovic, T. et al. (2009). Power control in two-tier femtocell networks. *IEEE Trans. Wireless Commun.* 8 (8): 4316–4328.

67 Yavuz, M., Meshkati, F., Nanda, S. et al. (2009). Interference management and performance analysis of UMTS/HSPA+femtocells. *IEEE Commun. Mag.* 47 (9): 102–109.

68 Femto Forum. http://www.femtoforum.org/femto/. (accessed 7 May 2020).

69 Blaunstein, N.Sh. and Sergeev, M.B. (2012). Definition of the channel capacity for femto-macrocell employment in urban environment with high layout of users. *J. Inf. Control Syst.* 3 (58): 54–62.

70 Tsalolihin, E., Bilik, I., Blaunstein, N., and Babich, Y. (2012). Channel capacity in mobile broadband heterogeneous networks based on femto cells. *Proceedings of EuCAP-2012 International Conference*, Prague, Czech Republic (26–30 March 2012), pp. 1–5.

71 Blaunstein, N. and Levin, M. (1996). VHF/UHF wave attenuation in a city with regularly spaced buildings. *Radio Sci.* 31 (2): 313–323.

72 Blaunstein, N. (1999). Prediction of cellular characteristics for various urban environments. *J. Anten. Propag. Mag.* 41 (6): 135–145.

73 Blaunstein, N. (1998). Average field attenuation in the non-regular impedance street waveguide. *IEEE Trans. Anten. Propag.* 46 (12): 1782–1789.

74 Blaunstein, N., Katz, D., Censor, D. et al. (2002). Prediction of loss characteristics in built-up areas with various buildings' overlay profiles. *J. Anten. Propag. Mag.* 44 (1):181–192.

75 Yarkoni, N., Blaunstein, N., and Katz, D. (2007). Link budget and radio coverage design for various multipath urban communication links. *Radio Sci.* 42 (2): 412–427.

76 Ben-Shimol, Y., Blaunstein, N., and Sergeev, M.B. (2015). Depolarization effects in various built-up environments. *Sci. J. Inf. Control Syst.* 69 (2): 83–94.

77 Okumura, Y., Ohmori, E., Kawano, T., and Fukuda, K. (1968). Field strength and its variability in the VHF and UHF land mobile radio service. *Rev. Electric. Commun. Lab.* 16 (9–10): 825–843.

78 Wells, P.J. (1977). The attenuation of UHF radio signal by houses. *IEEE Trans. Veh. Technol.* 26 (4):358–362.

79 Bertoni, H.L. (2000). *Radio Propagation for Modern Wireless Systems.* Upper Saddle River, NJ: Prentice Hall PTR.

80 Seidel, S.Y. and Rappaport, T.S. (1992). 914 MHz path loss prediction models for indoor wireless communications in multifloored buildings. *IEEE Trans. Anten. Propag.* 40 (2): 200–217.

81 Yarkoni, N. and Blaunstein, N. (2006). Prediction of propagation characteristics in indoor radio communication environments. *J. Electromagn. Waves Appl.: Progr. Electromagn. Res. PIER* 59: 151–174.

82 Yu, W., Ginis, G., and Cioffi, J.M. (2002). Distributed multiuser power control for digital subscriber lines. *IEEE J. Sel. Areas Commun.* 20 (5): 1105–1115.

83 Scutari, G. and Barbarossa, D.P.P. (2008). Optimal linear precoding strategies for wideband non-cooperative systems based on game theory-part II: algorithms. *IEEE Trans. Signal Process.* 56 (3): 1250–1267.

84 Scutari, G., Palomar, D.P., and Barbarossa, S. (2008). Asynchronous iterative water-filling for Gaussian frequency-selective interface channels. *IEEE Trans. Inf. Theory* 54 (7): 2868–2878.

85 WiMAX Forum (2008). WiMAX system evaluation methodology, V2-1, 230 pp.

86 Eklund, K., Marks, R.B., Kenneth, L., and Wang, S. (2002). IEEE standard 802.16: a technical overview on the WirelessMAN air interface for broadband wireless access. *IEEE Commun. Mag.* 40: 98–107.

87 Sengupta, S., Chatterjee, M., and Ganguly, S. (2008). Improving quality of VoIP streams over WiMAX. *IEEE Trans. Comput.* 57: 145–156.

88 So-In, C., Jain, R., and Tamimi, A.K. (2009). Scheduling in IEEE 802.16e mobile WiMAX networks: key issues and a survey. *IEEE J. Sel. Areas Commun.* 27(2): 156–171.

89 Niyato, D. and Hossain, E. (2006). Queue-aware uplink bandwidth allocation and rate control for polling service in IEEE 802.16 Broadband wireless networks. *IEEE Trans. Mob. Comput.* 5 (6): 668–679.

90 Cicconetti, C., Erta, A., Lenzini, L., and Mingozzi, E. (2007). Performance evaluation of the IEEE 802.16 MAC for QoS support. *IEEE Trans. Mob. Comput.* 6 (1): 26–38.

91 Taahol, P., Salkintzis, A.K., and Iyer, J. (2008). Seamless integration of mobile WiMAX in 3GPP networks. *IEEE Commun. Mag.* 46:74–85.

92 Etemad, K. (2008). Overview of mobile WiMAX technology and evolution. *IEEE Commun. Mag.* 46: 31–40.

93 Niyato, D. and Hossain, E. (2007). Integration of WiMAX and WiFi: optimal pricing for bandwidth sharing. *IEEE Commun. Mag.* 45 (5): 140–146.

94 Nie, J., Wen, J., Dong, O., and Zhou, Z. (2007). A seamless handoff in IEEE 802.16a and IEEE 802.11 hybrid networks. *Proceedings of International Conference on Convergence Information Technology*, pp. 24–29.

95 Nie, J., He, X., Zhou, Z., and Zhou, C. (2005). Communication with bandwidth optimization in IEEE 802.16 and IEEE 802.11 hybrid networks. *Proceedings International Symposium on Communications and Information Technologies*, pp. 26–29.

96 Kinoshita, K., Yoshimoto, M., Murakami, K., and Kawano, K. (2010). An effective spectrum sharing method for WiFi/WiMAX interworking mesh network. *Proceedings of IEEE Conference WCNC*, pp. 986–990.

97 Kinoshita, K., Kanamori, Y., Kawano, K., and Murakami, K. (2010). A Dynamic spectrum assignment method based on call blocking probability prediction in WiFi/WiMAX integrated networks. *Proceedings of IEEE Conference WCNC*, pp. 991–995.

98 Yang, W.-H., Wang, Y.-Ch., Tseng, Yu-Ch., and Lin, B.-Sh.P. (2010). An energy-efficient handover scheme with geographic mobility awareness in WiMAX-WiFi integrated networks. *Proceedings of IEEE Conference WCNC*, pp. 996–1001.

99 3GPP (2005). UTRA-UTRAN long term evolution (LTE) and 3GPP system architecture. http://www.3gpp.org/article/lte (accessed 7 May 2020).

100 Zyren, J. and McCoy, W. (2007).' Overview of the 3GPP long term evolution physical layer. 3G Americas white paper, Doc. Number: 3GPPEVO-LUTIONWP, July 2007.

101 3GPP. TR 36.912 (2009). Feasibility Study for Further Advancements of E-UTRA (LTE-Advanced). http://www.3gpp.org/article/lte-advanced (accessed 7 May 2020).

102 3GPP. TR 36.913 (2010). Requirements for Further Advancements for Evolved Universal Terrestrial Radio Access (E-UTRA) (LTE-Advanced). http://www.3gpp.org/article/lte-advanced (accessed 7 May 2020).

103 Ghosh, A., Ratasuk, R., Mondal, B. et al. (2010). LTE-advanced: next generation wireless broadband technology. *IEEE Wireless Commun.* 17 (3): 10–22.

104 Parkvall, S., Furuskar, A., Dahlman, E., and Research, E. (2011). Evolution of LTE toward IMT-advanced. *IEEE Commun. Mag.* 50 (5): 84–91.

105 Astey, D., Dahlman, E., Furuskar, A. et al. (2009). LTE: the evolution of mobile broadband. *IEEE Commun. Mag.* 47 (2): 44–51.

106 3GPP Technical Specification Group Radio Access Network (2010). Evolved Universal Terrestrial Radio Access (E-UTRA), Physical Layer Procedures (Release 9) 3GPP TS36.213 V9.3.0, June 2010.

107 3GPP Technical Specification Group Radio Access Network (2010). Evolved Universal Terrestrial Radio Access (E-UTRA), Further advancements for E-UTRA physical layer aspects (Release 9) 3GPPTS36.814V9.0.0, March 2010.

108 Ghaffar, R. and Knopp, R. (2011). Interference-aware receiver structure for Multi-User MIMO and LTE. *EURASIP J. Wireless Commun. Netw.* 40:24.

109 Li, Q., Li, G., Lee, W. et al. (2010). MIMO techniques in WiMAX and LTE: a future survey. *IEEE Commun. Mag.* 48 (5): 86–92.

110 Kusume, K., Dietl, G., Abe, T. et al. (2010). System level performance of downlink MU-MIMO transmission for 3GPP LTE-advanced 2010. *IEEE 71st Vehicular Technology Conference*, Taipei, pp. 1–5.

111 Covacacs, I.Z., Ordonez, L.G., Navarro, M. et al. (2010). Toward a reconfigurable MIMO downlink air interface and radio resource management: the SURFACE concept. *IEEE Commun. Mag.* 48 (6): 22–29.

112 EU FP7 Project SAMURAI – Spectrum Aggression and Multi-User MIMO: Real-World Impact. http://www.ict‐samurai.eu/page1001 .en.htm.

113 3GPP TSG RAN WG1 #62 (2010). Way Forward on Transmission Mode and DCI design for Rel-10 Enhanced Multiple Antenna Transmission R1-105057, Madrid, Spain (August 2010).

114 3GPP TSG RAN WG1 #62 (2010). Way Forward on 8 Tx Codebook for Release 10 DL MIMO R1-105011, Madrid, Spain (August 2010).

115 3GPP TR 36.942 V10.3.0 (2012-06), 3rd Generation Partnership Project (2012). Technical Specification Group Radio Access Network; Evolved Universal Terrestrial Radio Access (E-UTRA); Radio Frequency (RF) system scenarios (Release 10) (June 2012).

116 3G Americas white paper (2010). 3GPP mobile broadband innovation path to 4G: Release 9, Release 10 and beyond: HSPA+, SAE/LTE and LTE-advanced. http://www.4gamericas.org/documents/3GPP_Rel-9_Beyond %20Feb%202010.pdf (accessed 7 May 2010).

117 Duplicy, J., Badic, B., Balraj, R. et al. (2011). MU-MIMO in LTE systems. *EURASIP J. Wireless Commun. Netw.* Article ID 496763, 13 pages. https://doi.org/10.1155/2011/496763.

118 3GPP TR 36.942 V8.4.0 (2012-06) (2012). 3rd Generation Partnership Project; Technical Specification Group Radio Access Network; Evolved Universal Terrestrial Radio Access (E-UTRA); Radio Frequency (RF) system scenarios (Release 8) (June 2012).

119 Zhang, H., Prasad, N., and Rangarajan, S. (2011). MIMO Downlink Scheduling in LTE and LTE-Advanced Systems. Tech. Rep. NEC Labs America. http://www.nec-labs.com/~honghai/TR/lte-scheduling.pdf (accessed 7 May 2010).

Part II

Physical Layer of Wireless Networks Beyond 4G

3

Link Budget Design in Terrestrial Communication Networks

A link budget is a set of calculations aimed at estimating the power of the signal recorded by the receiver and the noise power that affected the receiver. The received power is estimated by knowing the transmitted power, the separation distance between terminal antennas, propagation conditions, frequency, and the terminal antenna gains [1–9]. From the received power, another important parameter of the communication, such as the ratio of energy of one information bit to noise power E_b/N_0, can be estimated, which directly determines the achieved bit error rate (BER) of the communication channel. Furthermore, not only path loss and noise affect the link, other parameters such as channel fading, electronic system, and antenna losses must also be considered for more accurate prediction of the total path loss [10–15].

3.1 Total Path Loss and Link Budget – Physical Layer of Any Network

Wireless communication links include several channels having different physical principles and processes with their own independent or correlated working characteristics and operating elements. A simple scheme of such a link consists of a transmitter, a receiver, and the propagation channel.

A link budget (see Figure 3.1) accounts for all the gains and losses from the transmitter, through the medium (free space, cable, waveguide, etc.), to the receiver in a communication system. It accounts for the attenuation of the transmitted signal due to propagation.

To summarize total link budget, we use [9, 15]

$$L_{\text{tot}} = N_0 + A_t + \overline{L} + G_T + G_R + L_{\text{SF}} + L_{\text{FF}} \text{ (dB)} \tag{3.1}$$

$$\overline{L} = L_{\text{LOS}} + L_{\text{NLOS}} \tag{3.2}$$

Advanced Technologies and Wireless Networks Beyond 4G, First Edition.
Nathan Blaunstein and Yehuda Ben-Shimol.
© 2021 John Wiley & Sons, Inc. Published 2021 by John Wiley & Sons, Inc.

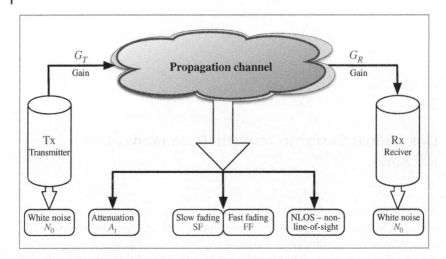

Figure 3.1 Link budget for any wireless communication network

3.1.1 White Noise

This noise occurs in the electronic devices and in the antenna of the transmitter and the receiver due to thermal movements of electrons via the electronic elements and devices. In other words, thermal noise is created by random motion of the electrons within the electronic components, when temperature exceeds absolute temperature of background [9]

$$T_0 = 290 \text{ K} = 17 \text{ °C}$$
$$N_0 = T_0 \cdot k_B \cdot B_w \cdot F \tag{3.3}$$

where $k_B = 1.38 \times 10^{-23}$ $(\frac{\text{W s}}{\text{K}})$ is the Boltzmann constant, B_w is the bandwidth of the wireless system, $F = T_0(1 + T_e/T_0)$ is the noise factor, and T_e is the effective noise temperature.

3.1.2 Slow Fading

Slow fading is used by events such as shadowing, where a large obstruction such as a hill or a large building obscures the main path between the transmitter and the receiver. The received power change caused by shadowing is often modeled using a log-normal distribution or Gaussian distribution (Figure 3.2) with a standard deviation according to log-distance path loss or distance path loss. The probability density function (PDF) of slow fading [1, 5–9] is

$$\text{PDF}(r) = \frac{1}{\sigma_L \sqrt{2}} \exp\left[-\frac{(r - \bar{r})^2}{2\sigma_L^2}\right] \tag{3.4}$$

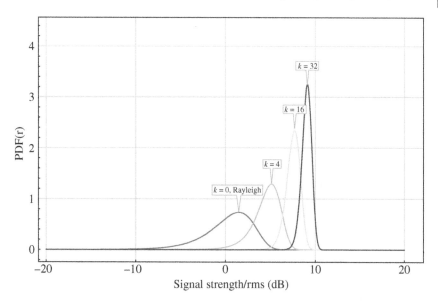

Figure 3.2 Fast-fading description via the Ricean PDF

3.1.3 Fast Fading

In the case of receiver or transmitter movement, the spatial variations of the resultant signal at the receiver are temporal. The received signal by the mobile at any spatial point may consist of many signals having randomly distributed amplitudes, phases and angles of arrival, and different time delays (see Figure 3.2). This temporal fading is associated with a shift of frequency radiated by the stationary receiver/transmitter. In a fast-fading channel, the channel impulse response changes rapidly within the symbol duration. To estimate the contribution of each signal component at the receiver, due to the dominant (or line of sight, LOS) and the secondary (or multipath), the Ricean parameter K of fast fading is usually introduced as a ratio between these components [1, 5–9]:

$$K = \frac{\text{LOS component power}}{\textit{Multipath} \text{ component power}} \tag{3.5}$$

The K parameter can be rewritten using the following formula [1, 5–9]:

$$K = \frac{A^2}{2\sigma^2} \tag{3.6}$$

Using the above defined parameter, we can write the PDF function of fast fading [1–9]:

$$\text{PDF}(r) = \frac{r}{\sigma^2} e^{\left(-\frac{r^2}{2\sigma^2}\right)} e^{-K} I_0 \left(\frac{r}{\sigma}\sqrt{2K}\right) \tag{3.7}$$

From Figure 3.2 it is clear that increasing the Ricean K parameter of fast fading from 4 to 32, i.e. with $K \gg 1$, the Rayleigh PDF limits to a Dirac-shaped function.

3.1.4 Antenna Gain

The gain of an antenna is closely associated with directivity. Directivity is the ability of an antenna to focus energy in a specific direction when transmitting or receiving energy better from a specific direction as seen in Figure 3.3.

A gain is defined as a relative measure of an antenna's ability to direct or concentrate radio-frequency energy at a specific direction. It is typically measured in decibel (dB). In a receiving antenna, the gain describes how well the antenna converts radio waves arriving from a specific direction into electrical power [9, 15]:

$$G = 10 \log_{10}(g) \ (dB) \tag{3.8}$$

3.1.5 Average Attenuation

The average attenuation $\bar{L}$, defined in Eq. (3.2), consists of two factors or components [9, 15]:

- LOS or line of sight.
- NLOS or non-line-of-sight.

3.1.5.1 Line of sight
LOS illustrates a type of propagation that can transmit and receive power only where the transmitter and the receiver are in view of each other without any kind of obstacle between them. Long-distance data communication is more

(a) (b)

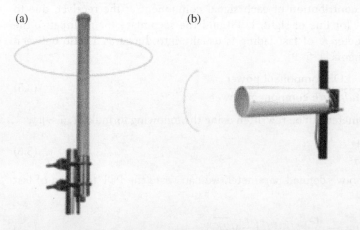

Figure 3.3 The 360° spatial coverage (a) and the focused spatial coverage (b)

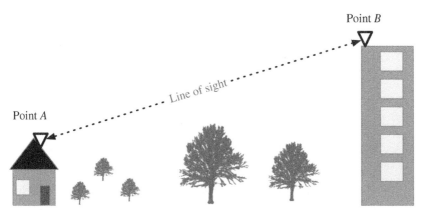

Figure 3.4 Line-of-sight (LOS) propagation link

effective through wireless networks (see Figure 3.4), but geographical obstacles and the curvature of the earth bring limitation to line-of-sight transmission.

In fact, RF signals do not propagate along straight lines. Due to the refractive effects of atmospheric layers, the propagation path is somewhat curved. Thus, the maximum service range of the station is not equal to the line-of-sight geometric distance $d = |AB|$.

$$d \approx \sqrt{2kRh} \tag{3.9}$$

where k is a factor means geometrically reduced bulge and a longer service range, R represents the radius of Earth, and h represents the altitude of transmitting antenna (Tx).

Since the altitude of the station is much less than the radius of the Earth, it is defined as ($h \ll R$)

$$d \approx 4.12\sqrt{h}$$

3.1.5.2 Non-line-of-sight

NLOS is a type of propagation where transmitter and receiver are not in view of each other. Long-distance communication is not effective in far distances due to the reflection from the object that observed part of the energy.

As we attempt to obtain $\overline{L}$ which consists of L_{LOS} and L_{NLOS}, we understand that those parameters are dependent on many different situations such as open spaces, built-up areas, and forest areas. Due to these various situations, we need to consider what we desire from our communication system. All models describe specific scenarios observed in rural, suburban, and urban areas, and if we want one model to obtain the best quality in all situations, we must continue and search for the best solution. Figure 3.5 presents all the losses and the gain we achieve in every environment under test.

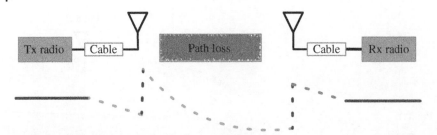

Figure 3.5 Total path loss in wireless communication link

Each model demonstrates a specific case and cannot be used everywhere. These models show us the calculation of the channel's path loss along the transmitted route. We also have an antenna gain that increases the received/transmitted signal and the cable losses.

3.2 The Terrain Propagating Models for Total Path Loss Prediction

There are many models created during the past 50 years to evaluate the average attenuation (or path loss) in various terrestrial environments. We present here only those attractive models the accuracy of which was proved by numerous experiments carried in various built-up terrain environments.

3.2.1 Hata–Okumura Model

Okumura created his own model based on a lot of measurements carried out in the city of Tokyo, Japan. This is an ideal large urban city with many urban structures but not with many tall blocking structures. This model is based on measured data and does not provide any analytical explanation, that is, a classical empirical model to measure the radio signal. Therefore, Okumura model is among the simplest and best in terms of accuracy in path loss prediction for mature cellular and land mobile radio systems in cluttered environments. It is very practical and has become a standard for system planning in modern land mobile radio systems in Japan. This model is applicable for frequencies in the range of 150–1920 MHz (although it is typically extrapolated up to 3000 MHz) and distances of 1–100 km. It can be used for base-station antenna heights ranging from 30 to 1000 m [9, 15].

$$L_{50\%} = L_F + A_{mu}(f, d) + G(h_{tb}) + G(h_{re}) - G_{area} \tag{3.10}$$

where

$L_{50\%}$ is the 50th percentile (median) value of propagation path loss.

L_F is the free-space propagation loss.

$A_{mu}(f, d)$ is the median attenuation relative to free space.

$G(h_{he})$ is the base-station antenna height gain factor.

$G(h_{re})$ is the mobile antenna height gain factor.

G_{area} is the gain due to the type of environment.

Furthermore, Okumura found that % Preview source code for paragraph 0

$$
\begin{aligned}
G(h_{te}) &= 20 \log_{10}\left(\frac{h_{te}}{200}\right) & 30 \text{ m} < h_{te} < 1000 \text{ m} \\
G(h_{re}) &= 10 \log_{10}\left(\frac{h_{re}}{3}\right) & h_{re} < 3 \text{ m} \\
G(h_{re}) &= 20 \log_{10}\left(\frac{h_{re}}{3}\right) & 3 \text{ m} < h_{re} < 10 \text{ m}
\end{aligned}
\tag{3.11}
$$

The major disadvantage with this model is its slow response to rapid changes in terrain; therefore, the model is good in urban and suburban areas but not as good in rural areas.

Then Hata, based on the empirical data obtained by Okumura, had created a so-called semiempirical model that now in the literature is called the Hata–Okumura model. Simply speaking, this is an empirical formulation of the graphical path loss data provided by Okumura and is valid from 150 to 1500 MHz. The predictions of the Hata model are very similar to the original Okumura model if a distance between terminal antennas, d, exceeds 1 km. Hata presented the urban formula as a standard formula and supplied correction equations for application to other built-up scenarios. Thus, for urban areas, the standard formula in dB is

$$
L(\text{urban}) = L_{50}(\text{urban}) + 10\gamma \log_{10} d \tag{3.12}
$$

where

$$
L_{50}(\text{urban}) = 69.55 + 26.16 \log_{10}(f_c) - 13.82 \log_{10}(h_{te}) - a(h_{re}) \tag{3.13}
$$

$$
\gamma = (44.9 - 6.55 \log(h_{te}))/10 \tag{3.14}
$$

with the following definitions:

- f_c(MHz) is the frequency.
- $h_{te}(m)$ is the base-station height ranging from 30 to 200 m.
- $h_{re}(m)$ is the receiver's antenna height ranging from 1 to 10 m.
- $d(m)$ is the distance between the transmitter and the receiver.

For a large city:

$$a(h_{re}) = \begin{cases} 8.29(\log 1.54 h_{re})^2 - 1.1 \text{ dB} & f_c \leq 300 \text{ MHz} \\ 3.2(\log 11.75 h_{re})^2 - 4.97 \text{ dB} & f_c \geq 300 \text{ MHz} \end{cases} \tag{3.15}$$

For suburban areas:

$$L_{50} = L_{50}(\text{urban}) - 2\left[\log\left(\frac{f_c}{28}\right)\right]^2 - 5.4 \tag{3.16}$$

and for free space:

$$L_{50} = L_{50}(\text{urban}) - 4.78(\log f_c)^2 - 18.33 \log f_c - 40.98 \tag{3.17}$$

Later, the Hata model was standardized and extended the urban Hata–Okumura model to cover a more elaborated range of frequencies. In the literature this model is also called the "Hata Model PCS Extension."

$$L_{50}(\text{urban}) = 46.3 + 33.9 \log(f_c) - 13.82 \log(h_{te}) - a(h_{re}) +$$
$$(44.9 - 6.55 \log(h_{te})) \log(d) + C_M \tag{3.18}$$

where $C_M = 0$ dB for medium cities and suburban areas, and $C_M = 3$ dB for metropolitan centers.

3.2.2 Bertoni Multidiffraction Model

This model is most suitable for flat suburban and urban areas with uniform building height. Among other models, this model gives a more precise average path loss prediction proved by numerous experiments carried in the United States [4]. This is a result of additional parameters introduced, which characterize different environments.

Figure 3.6 shows a uniform height building with equal separation, which is the core feature of the Walfisch–Bertoni model. The main concern is if the building has a low roof top, it should be the route for analysis using Bertoni's model that considers buildings with lower height uniform. This analysis is based on

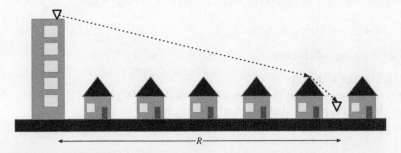

Figure 3.6 Plot of Walfisch–Bertoni model

a ray-tracing technique, which considers each building block as a diffraction screen.

The path loss equation for LOS conditions is given by

$$PL_{LOS} = 42.6 + 26\log(d) + 20\log(f) \tag{3.19}$$

where d is the separation (in km) between the transmitter and the receiver, and f is the frequency (in MHz).

For NLOS conditions:

$$PL_{NLOS} = L_{FSL} + L_{rts} + L_{msd} + L_{FS} \text{ if } L_{rts} + L_{msd} > 0 \tag{3.20}$$

where

- L_{FSL} is the free-space loss.
- L_{rts} is the rooftop to street diffraction.
- L_{msd} is the multiscreen diffraction loss.

3.2.3 Walfisch–Ikegami Model (COST 231 Standard) Based on Analytical Bertoni Model

This model was created for microcells and small macrocells by combining the analytical models of Walfisch and Bertoni with the empirical model of Ikegami. This combined model distinguishes between the LOS and NLOS cases. The average height of the buildings and the average spacing value imply terrain that is more suitable for suburban areas. The accuracy in this case is high because in urban environments the propagation in the vertical plane and over the rooftops (multiple diffractions) is dominating, especially if the transmitters are mounted above roof top levels (see Figure 3.7).

For LOS, the total path loss is

$$PL_{LOS} \text{ (dB)} = 42.6 + 26\log(d) + 20\log(f) \tag{3.21}$$

and for NLOS conditions, the total path loss is

$$P_{NLOS} = \begin{cases} L_{FSL} + L_{rts} + L_{msd} + L_{FS} & L_{rts} + L_{msd} > 0 \\ L_0 & L_{rts} + L_{msd} < 0 \end{cases} \tag{3.22}$$

where

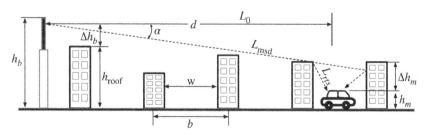

Figure 3.7 Walfisch–Ikegami model geometrical plot

- L_0 is the free-spacepath loss.
- L_{msd} is the multiscreen loss along the propagation path.
- L_{rts} is the rooftop to street diffraction and scatter loss.

We can find multiscreen loss along the propagation path via the following formula:

$$L_{\text{msd}} = L_{bsh} + k_a + k_d \log(d) + k_f \log(f_c) - 9\log(b) \tag{3.23}$$

where b is the distance between two adjacent buildings and

$$L_{bsh} = 18 + \log(1 + \Delta h_b) \quad \text{for } h_b > h_{\text{roof}} \tag{3.24}$$

$$k_a = \begin{cases} 54 & h_b > h_{\text{roof}} \\ 54 - 0.8\Delta h_b & d \geq 0.5 \ \text{km} \wedge h_b \leq h_{\text{roof}} \\ 54 - 1.6\Delta h_b & d < 0.5 \ \text{km} \wedge h_b = h_{\text{roof}} \end{cases} \tag{3.25}$$

$$k_d = \begin{cases} 18 & h_b > h_{\text{roof}} \\ 18 - \dfrac{15\Delta h_b}{h_{\text{roof}}} & h_b \leq h_{\text{roof}} \end{cases} \tag{3.26}$$

$$k_f = \begin{cases} 0.7\left(\dfrac{f_c}{925} - 1\right) & \text{for medium-sized cities} \\ 1.5\left(\dfrac{f_c}{925} - 1\right) & \text{for large-sized cities} \end{cases} \tag{3.27}$$

We also need to consider L_{rts}:

$$L_{\text{rts}} = -16.9 + -10\log w + 10\log f_c + 20\log \Delta h_m + L_{\text{ori}} \tag{3.28}$$

where, as follows from Figure 3.7, w is the width of the street, and Δh_m is the difference between the building height and the height of the received antenna

$$L_{\text{ori}} = \begin{cases} -10 + 0.354\varphi & 0° \leq \varphi \leq 35° \\ 2.5 + 0.075(\varphi - 35°) & 35° \leq \varphi \leq 55° \\ 4 + 0.114(\varphi - 55°) & 55° \leq \varphi \leq 90° \end{cases} \tag{3.29}$$

3.2.4 Stochastic Multiparametric Model

3.2.4.1 Parameters of the model

This model considers an array of buildings that are randomly distributed on a rough terrain [9–15]. To obtain a predictor of the average path loss in urban and suburban areas, as in [10–15], we must account the main features of the built-up terrain such as the density of buildings, v ($1/\text{km}^2$), and $\overline{L}$ – the average length of buildings (or the width, depending on the orientation with respect to the direction of the antenna). So, we put our attention on the parameter of the 1-D density of the buildings' contours [9–15]

$$\gamma_0 = \frac{2\bar{L}v}{\pi} \tag{3.30}$$

3.2.4.2 Effect of buildings' overlap profile

To understand the influence of the built-up area relief on signal intensity, we introduce, following [9–15], the average building height

$$h = h_2 - \frac{n(h_2 - h_1)}{n+1} \tag{3.31}$$

and the profile can be presented as

$$P_h(z) = H(h_1 - z) + H(z - h_1)H(h_2 - z)\left[\frac{h_z - z}{h_2 - h_1}\right]^n \tag{3.32}$$

Here, $H(\cdot)$ is the Heaviside step function, n is a parameter of the built-up profile, h_1 is the minimal building height, and h_2 is the maximum building height.

In Figure 3.8, $n \gg 1$ describes a case where the buildings' heights are close to the lowest buildings of height h_1 in the tested area.

$n \ll 1$ describes a case where the buildings' heights are close to the tallest buildings of height h_2 in the tested area. Finally, when $n \approx 1$, the numbers of low and tall buildings are approximately the same.

Several geometrical factors of the built-up layer, which should also be taken into account and will be shown in our further discussions in Chapters 4 and 5, are the terminal antennas' heights with respect to the maximum and minimum building heights by introducing the so-called profile function $F(z_1, z_2)$ between

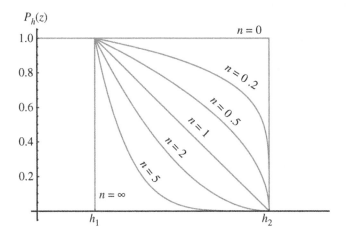

Figure 3.8 Graph of $P_h(z)$ vs. the height z of buildings varied from minimum and maximum building heights

two antennas located at heights z_1 and z_2:

$$F(z_1, z_2) = \int_{z_1}^{z_2} P_h(z)dz \tag{3.33}$$

By inspecting the displayed curves in Figure 3.8, it is clear that for a constant transmitter antenna height, there is an increase in the height of the receiver antenna, the value of $F(z_1, z_2)$ becomes smaller, and the effect of the building layer on the path loss is reduced, as analytically and numerically was shown in [16, 17].

Returning to the multiparametric stochastic approach described earlier, and using expressions (3.32) and (3.33), describing built-up terrain profile description, one can follow the corresponding algorithm described in [9, 15] to obtain the probability of fading and its impact in the total path loss due to diffraction phenomenon from buildings and other obstructions located in the areas of service. Thus, following [9, 15], we can obtain a full probability, called the complimentary cumulative distribution function

$$\text{CCDF}(z_1, z_2, n) = \frac{1}{z_2 - z_1} \int_{z_1}^{z_2} P_h(z)dz \equiv \frac{1}{z_2 - z_1} F(z_1, z_2). \tag{3.34}$$

Then, in dB, the influence of slow fading can be evaluated by $L = \log A$, where

$$A = \frac{1}{z_2 - z_1} F(z_1, z_2) \tag{3.35}$$

3.2.4.3 Signal intensity distribution

In built-up areas such as mixed residential or suburban, the effects of the overlay profile of the buildings are not so actual, since all the obstructions such as houses and trees are approximately of the same height. Under such conditions, we can finally get the following formula for the incoherent part of the total field intensity [9–15]:

$$\langle I_{\text{inc}} \rangle = \frac{\Gamma}{8\pi} \frac{\lambda \ell_h}{\lambda^2 + (2\pi \ell_h \gamma_0)^2} \times$$
$$\frac{\lambda \ell_v}{\lambda^2 + [2\pi \ell_v \gamma_0 (h - z_1)]^2 d^3} \left[\left(\frac{\lambda d}{4\pi^3} \right)^2 + (z_2 - h)^2 \right]^{1/2} \tag{3.36}$$

This formula contains two dimensions of the obstructions in both directions – vertical and horizontal.

The coherent part of the total field intensity can be written according to [9–15] as

$$\langle I_{\text{co}} \rangle = \exp\left(-\gamma_0 d \frac{h - z_1}{z_2 - z_1} \right) \left[\frac{\sin\left(\frac{k z_1 z_2}{d} \right)}{2\pi d} \right]^2 \tag{3.37}$$

where

- ℓ_h is the scale of coherency in horizontal plane.
- ℓ_v is the scale of coherency in vertical plane.
- h is the average building height.
- d is the distance between the terminal points z_1 and z_2.

The principal difference between suburban or mixed residential areas and urban environments is that in urban areas the building dimensions are much larger than the wavelength. In this case, we can exclude the influence of the reflecting properties of the walls of the buildings along the horizontal axis. This allows us to deduce the 3D case to 2D case, that is,

$$\langle I_{\text{inc1}} \rangle = \frac{\Gamma}{8\pi} \frac{\lambda \ell_v}{\lambda^2 + [2\pi \ell_v \gamma_0 F(z_1, z_2)]^2 d^3} \left[\left(\frac{\lambda d}{4\pi^3} \right)^2 + (z_2 - h)^2 \right]^{1/2} \tag{3.38}$$

The corresponding formula for double scattering and diffraction is given by

$$\langle I_{\text{inc2}} \rangle = \frac{\Gamma^2 \lambda^3 \ell_v^2}{24\pi^2 \{ \lambda^2 + [2\pi \ell_v \gamma_0 F(z_1, z_2)]^2 \}^2 d^3} \left[\left(\frac{\lambda d}{4\pi^3} \right)^2 + (z_2 - h)^2 \right]$$

$$\tag{3.39}$$

The coherent part of the total field intensity can be obtained:

$$\langle I_{\text{co}} \rangle = \exp\left(-\gamma_0 d \frac{1}{z_2 - z_1} \int_{z_1}^{z_2} P_h(z) dz \right) \left[\frac{\sin\left(\frac{k z_1 z_2}{d} \right)}{2\pi d} \right]^2 \tag{3.40}$$

$$\langle I_{\text{total}} \rangle = \langle I_{\text{co}} \rangle + \langle I_{\text{inc1}} \rangle + \langle I_{\text{inc2}} \rangle \tag{3.41}$$

3.3 Validation of Most Suitable Models via the Recent Experiments

We check the accuracy of these three more attractive and usually used models in networks beyond 4-G using the experimental data obtained by group of researchers from Ben-Gurion University (BGU), led by the authors of this book. The examined models are Hata-Okumura semiempirical model, the Walfisch–Ikegami semiempirical model, the Bertoni physical model, and the stochastic multiparametric model [9, 15].

The first campaign was carried out in Jerusalem. The data from the topographic map of Jerusalem is presented in Table 3.1. Figure 3.9 presents the comparison of Hata, Walfisch–Ikegami, and the multiparametric model path loss prediction vs. experimental data along the distance between the stationary Tx antenna and moving Rx antenna.

Table 3.1 Average parameters of the city of Jerusalem

v (km^{-2})	$\bar{L}_{avg}$ (m)	f (MHz)	h_T (m)	h_R (m)	γ_0 (km^{-1})
103.9	18	930	42	1.5	1.27

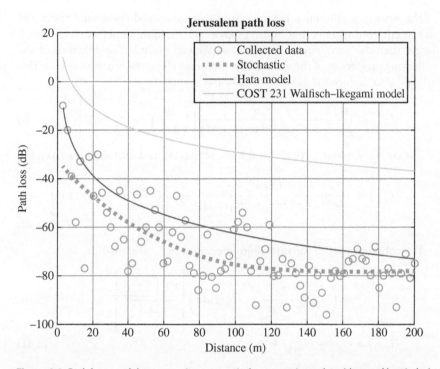

Figure 3.9 Path loss models vs. experiments carried out over Jerusalem (denoted by circles)

We also achieved the errors of each model with respect to the experimental data. The lowest standard deviation around measured data was given by the stochastic model, $\sigma = 3.4463$ dB, whereas from the Hata model, we got $\sigma = 31.5725$ dB. The Walfisch–Ikegami model was ignored due to its irrelevancy (see the top curve which is too far from the experimental data). In the distance range of 12–100 m, the stochastic model results with $\sigma = 2.1113$ dB, whereas the Hata model results with $\sigma = 5.4478$ dB from the experimental data. Note that due to Jerusalem's topographic map, the density profile was changed according to definitions of γ_0 and γ_{12}. These parameters are allowing us to create 2D map with vertical and horizontal planes $P_{\text{horizontal}} \sim e^{-\gamma_0 z}$ and $P\{z_1, z_2\}$, where again, z_1 and z_2 are the MS and BS heights, respectively, and

$$P[z_1, z_2] = \int_{h_{\min}}^{h_{\max}} P_h(h)dh \tag{3.42}$$

As a result, we can state that in the case of Jerusalem with irregular built-up terrain, consisting of hills and buildings lying above them, the stochastic model is more accurate than the Hata model, whereas other models, such as Bertoni and Walfisch–Ikegami, are not relevant. We presented in Figure 3.9 only one area of Jerusalem. A full analysis based on its topographic map showed that the standard deviation of the stochastic and Hata models gives cumulative effect at the ranges of $\sigma = 2.1 \sim 3.45$ dB and $\sigma = 5.45 \sim 31.57$ dB, respectively.

The next campaign was carried out in Ramat-Gan area, a suburban area of Tel Aviv, having regular smooth terrain with regularly distributed rows of streets and random distribution of buildings lining each street.

Table 3.2 shows the main parameters of the built-up terrain with all features of the tested area around the stock market of RamatGan. The results of comparative analysis using the stochastic, Hata, and Bertoni–Walfisch models vs. the experimental data (denoted by circles) are plotted in Figure 3.10.

As follows from the comparative analysis presented in Figure 3.10, the stochastic model results with the smallest standard deviation with respect to experimental data, around $\sigma = 10.0168$ dB, whereas the Bertoni–Walfisch model gives $\sigma = 19.0339$ dB. The Hata model is far from experimental data at the small ranges (less than 200–300 m). In distance range of 200–500 m, the stochastic model gives $\sigma = 10.6804$ dB, the Hata model $\sigma = 29.6056$ dB, and Bertoni–Walfisch model $\sigma = 12.213$ dB. At distances between BS and MS of 500–800 m, the stochastic model gives $\sigma = 3.3448$ dB, the Hata model $\sigma = 2.394$ dB, and Bertoni–Walfisch model $\sigma = 3.571$ dB.

Hence, using these results we can outline that for missed residential and suburban areas with smooth terrain and regularly distributed buildings and streets, the Bertoni–Walfisch model is a better predictor of the average path loss (or attenuation) compared to the Hata model and coincides with the stochastic model. At the same time, in urban built-up environments with rough terrain and irregular distribution of building heights and density layout, as in Jerusalem, both Bertoni–Walfisch and Hata models are not so relevant, and the stochastic model can be considered as a better predictor of the average path loss. As will be shown in Chapters 4, 5, and 8, the stochastic model is also a better predictor of fading phenomena and other main characteristics of wireless networks. Therefore, in our future discussions of the subject, the stochastic model will be used as an attractive "physical background" of main parameters of wireless networks performance for modern networks beyond 4-G.

Table 3.2 Parameters of the built-up terrain of the stock market area in the city of Ramat Gan

v (km^{-2})	$\bar{L}_{avg}$ (m)	f (MHz)	h_T (m)	h_R (m)	γ_0 (km^{-1})
103.9	18	900	50	2	1.27

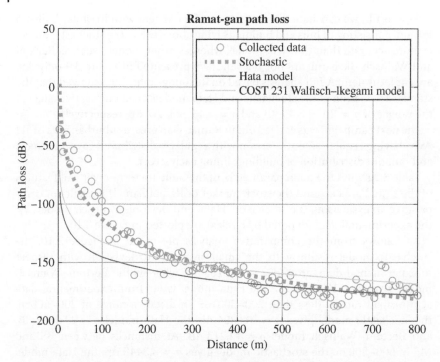

Figure 3.10 Path loss models vs. experiments carried out over RamatGan (denoted by circles)

3.4 Link Budget Design in Land–Atmosphere and Atmosphere–Land Communication Networks

During the recent decades, land-to-air and air-to-land communication links became most attractive for current and future wireless networks beyond 4-G due to the use of aircrafts, helicopters, drones, and so forth. As mentioned in [16–33], the troposphere consists of gaseous particles, called aerosols, rain particles, clouds, fog, hail, snow, etc.; all are usually called "hydrometeors" in the literature. These "hydrometeors" impact the attenuation, absorption, and scattering of radio waves passing the land–atmosphere channels. Furthermore, due to sporadic air streams and motions, another phenomenon – atmospheric turbulences – occurs in the troposphere. Due to irregular conditions in the troposphere, when a radio signal carrying data propagates through a tropospheric channel, its intensity varies sporadically. This phenomenon is called fading, fast and slow, in the time domain or small- and large-scale in the space domain, respectively. Fast fading plays an important role in link budget design and quality of service performance due to its huge possibility to obstruct signal with data passing such kinds of channels. Moreover, as will be shown below, the built-up terrain profile and density layout of ground-based obstructions (hills, trees, buildings, etc.) give significant effect on slow-fading phenomena and can decrease signal strength by 50%, causing huge loss of information data of more than 50%.

3.4.1 Content and Main Parameters of the Troposphere

The troposphere consists of different kinds of gaseous, liquid, and crystal structures such as gas molecules, aerosol, rain particles, cloud, fog, hail, and snow. All except the first are usually called hydrometeors in the literature. Furthermore, due to sporadic air streams and motions, another phenomenon occurs in space and time domains, called atmospheric turbulences. Below, we briefly describe the various components that make up the troposphere following the specific literature [16–33].

3.4.1.1 The content

Gaseous molecules and atoms There are many types of atmospheric molecules and atoms, such as O_2, O, CO_2, NO, and N_2 [29].

An aerosol For the purposes of this chapter, it is a system of liquid or solid particles uniformly dispersed in the atmosphere [9, 15, 27–29]. Aerosol particles play an important role in the perception process, providing the nuclei upon which condensation and freezing take place. The particles participate in chemical processes and influence the electrical properties of the atmosphere [20, 21, 27, 28]. The system begins to acquire the properties of a real aerosol structure when smaller particles are in suspension. An actual aerosol particle range can be between a few nanometers and about few micrometers, while aerosols composed of particles larger than 50 μm are unstable. The number of aerosol molecules can be found [20, 27, 28]:

$$N(z) = N(0) \exp\left(\frac{z}{z_s}\right) \tag{3.43}$$

where $N(0)$ is the number of molecules at the ground surface, z is the height of the molecules in meters, $N(z)$ is the current number of molecules, and z_s is the scale height while $1 \text{ km} < z_s < 1.4 \text{ km}$.

Clouds The shape, structure, and texture of clouds are influenced by air movements that change their formation and growth and are also influenced by the properties of the cloud particles themselves. There are four principal classes into which clouds are classified according to the kind of air motions that produce them [16, 22]:

(a) Layer clouds formed by the widespread regular ascent of air.
(b) Layer clouds formed by the widespread irregular stirring of turbulence.
(c) Cumuliform clouds formed by penetrative convection.
(d) Orographic clouds formed by ascent of air over hills and mountains.

In settled weather, clouds are small and well scattered. Their horizontal and vertical dimensions are only a kilometer or two. In disturbed weather, they cover a large part of the sky and can tower as high as 10 km or more. Clouds often cease their growth only upon reaching the stable stratosphere, producing heavy showers, hail, and thunderstorms. Growing clouds are sustained by upward air currents, which may vary in strength from a few centimeters per second to several meters per second. The effects of cloud on wave propagation in the troposphere are well known: scattering, absorption, and refraction, all of which cause attenuation and fading of the wave path. All these phenomena will be considered later.

Rain   Rain is the precipitation of liquid water drops with diameters greater than 0.5 mm [5, 17, 23–26]. When the drops are smaller, the precipitation is usually called drizzle. The concentration of raindrops typically spreads from 100 to 1000 m^{-3}. Drizzle droplets usually are more numerous. Raindrops seldom have diameters larger than 4 mm because the concentration generally decreases as the diameter increases, except when the rain is heavy. It does not reduce visibility as much as drizzle. Meteorologists classify rain according to its rate of fall. There are three classes of rain: light, moderate, and heavy and they correspond to dimensions less than 2.5 mm, between 2.8 and 7.6 mm, and more than 7.6 mm, respectively. Rain with rates of less than 250 mm per year and more than 1500 mm per year represents the extremes of rainfall for all the continents.

Atmospheric turbulence   This is a chaotic structure generated by irregular air movements in which the wind randomly varies in speed and direction [15, 18, 19]. Turbulence is important because it churns and mixes the atmosphere and causes water vapor, smoke, and other substances, as well as energy, to become distributed at all elevations. Atmospheric turbulence near the Earth's surface differs from that which occurs at higher levels. Within a few hundred meters of the surface, turbulence has a marked diurnal variation, reaching a maximum about midday. When the sky is cloudy, the low-level air temperature varies much less between day and night and the turbulence remains nearly constant. At altitudes of several thousand meters or more, the frictional effect of the Earth's surface topography on the wind is greatly reduced and the small-scale turbulence, which is usually observed in the lower atmosphere, is absent [30–33].

3.4.1.2   Main parameters of troposphere
The physical properties of the troposphere are characterized by the following main parameters such as temperature, T (in kelvin); pressure, P (in mbar or millimeters of mercury); and density, ρ (in particles per cubic meter or centimeter). All these parameters significantly change with altitude, season, and latitudinal variability and strongly depend on the weather [5, 18–21].

Temperature The temperature in the atmosphere depends on altitude h, in meters. The temperature T at height h (measured in meters) is given by [18–21]

$$T(h) = 288.15 + 0.065\ 45h \ \text{(K)} \tag{3.44}$$

The troposphere is a region between 10 and 20 km above the earth's surface, where the temperature is [18–21]

$$T(h) = 216.65 \ \text{(K)} \tag{3.45}$$

Pressure It is the force applied perpendicular to the surface of an object per unit area over which that force is distributed. The pressure can be determined by [18–21]

$$P(h) = 2.269 \times 10^4 \exp\left(-\frac{0.034\ 164(h - 11\ 000)}{216.65}\right) \ \text{(mbar)} \tag{3.46}$$

where h is the height (in m). In the troposphere, besides the atmospheric pressure, we usually need to know the water vapor partial pressure p, (in mbar) and the saturation pressure $e(t)$ [20, 21]. The relationship between water vapor pressure p_w and relative humidity is given by

$$p_w = \frac{\eta e_s}{100} \tag{3.47}$$

where

$$e_s = a\frac{bt}{t + c} \tag{3.48}$$

where η is the relative humidity (in %), t is the temperature (°C), e_s is the saturation pressure (pascal) for temperature t (°C), while the coefficients a, b, and c were defined empirically via numerous experiments [20, 21]. The vapor pressure p can be evaluated via the water vapor density ρ using the equation:

$$p_w = \frac{\rho_w T(h)}{216.7} \tag{3.49}$$

with the water vapor density ρ given by the following equation:

$$\rho = \rho_0 \exp\left(-\frac{h}{h_0}\right) \tag{3.50}$$

Here, h_0 is the scale height of 2 km, and the standard water vapor density is

$$\rho_0 = 7.5 \ \text{(g/m}^3) \tag{3.51}$$

Humidity In meteorology, the measurable quantity is the relative humidity $\eta(T)$, and we can relate p with $e_s(T)$. The relative humidity is given by [20, 21]

$$\eta = \frac{\rho_w}{e_s(T)} \tag{3.52}$$

3.4.2 Effects of Tropospheric Features on Signal Propagation

For the ideal fully gaseous atmosphere without hydrometeors, the fading phenomena of radio and optical waves can remain optimal 99.999% of the time for the paths of 5 km and more, with the fade margin of 28 dB. However, there are phenomena of propagation that can significantly decrease the efficiency of land-to-atmosphere or atmosphere-to-land communication links, such as scattering, attenuation, or absorption. Let us briefly consider the impact of each feature separately in the total path loss of the signal passing a tropospheric communication channel.

3.4.2.1 Main features occurring in the troposphere

Absorption Absorption (or attenuation) occurs because of conversion from wave energy to thermal energy within an attenuating particle, such as a gas molecule and different hydrometeors [9, 15].

Scattering Scattering is a vitally important feature causing strong fast fading of the signal and results from the redirection of the radio waves into various directions, so that only a fraction of the incident energy is transmitted onward in the direction of the receiver. The main scattering particles that are of interest to satellite systems are hydrometeors, including raindrops, fog, and clouds, and can be calculated using three main approaches that account for the relationship between the wavelength and the size of the particles causing the scatter. All approaches were discussed in [9, 15, 18, 30–33], and their scattering coefficients are described below by the following theoretical frameworks (3.53), (3.54) and (3.55):

(a) Mie scattering is applicable when the particle size is comparable to the radiation wavelength. The Mie scattering coefficient was defined as the ratio of the incident wave front that is affected by the particle to the cross-sectional [9, 15]:

$$\sigma_\lambda = \pi \int_{a_1} N(r)K(r,n)r^2 \, dr \tag{3.53}$$

(b) Rayleigh scattering applies when the radiation wavelength is much smaller than the particle sizes and is described in [9, 15] by

$$\sigma_\lambda = \frac{4\pi N V^2 (n^2 - n_0^2)^2}{\lambda^2 (n^2 + 2n_0^2)^2} \tag{3.54}$$

where n_0 is the index of refraction at the ground level.

(c) Nonselective scattering applies when the particle size is significantly larger than the radiation wavelength. Large-particle scattering is composed of contributions from three processes involved in the interaction of the electromagnetic radiation with the scattering particles [9, 15]:

$$\sigma_\lambda = \int N(r)Q(\lambda, m, r)\pi r^2 \, dr \tag{3.55}$$

where N is the number of particles per unit volume, V is the volume of scattering particles, K is a value between 0 and 4, r is the radius of spherical particle, n is an index of the refraction of waves at the layer of particles, and m is the mass of any particle.

The above formulas account for the following physical processes caused by gaseous particles of the troposphere, i.e.

- reflection from the surface of the particle with no penetration,
- passage through the particle with and without internal reflection, and
- diffraction at the edge of the particle.

Now, we examine separately the effects of each feature on signals passing the irregular troposphere, such as the effects of rain and cloud on the signal attenuation. The effect of turbulence causing scattering of signals will be considered separately as a main source of fast fading. There are three main causes for signal attenuations: molecular absorption, effects of rain, and effect of clouds. Then, the effect of turbulence on scattering of signals will be presented. We show the main parameters, the corresponding formulas, and compute and plot their characteristics in this paragraph.

3.4.2.2 Molecular–Gaseous absorption

Gaseous molecules found in the atmosphere may absorb energy from radio waves passing through them, thereby causing attenuation [9, 15, 29]. The signal degradation depends on frequency, temperature, pressure, and water vapor concentration and increases with them, as shown in Figure 3.11, calculated according to the equations below taken from ITU-676 standard (see Ref. [29]) for pressure $P = 1013$ mbar, temperature $T = 15\,°C$, and water vapor content $\rho_w = 7.5$ g/m^3. The following formulas have been used for computation of dependences shown in Figure 3.11.

The absorption in the atmosphere over path length r is given by [9, 15, 29]

$$A = \int \gamma(r) dr \tag{3.56}$$

where $\gamma(r)$ is a specific attenuation (in dB/ km) consisting of the sum of two components $\gamma_0(r)$ and $\gamma_w(r)$, the attenuation of oxygen and water vapor, respectively:

$$\gamma(r) = \gamma_o(r) + \gamma_w(r) \tag{3.57}$$

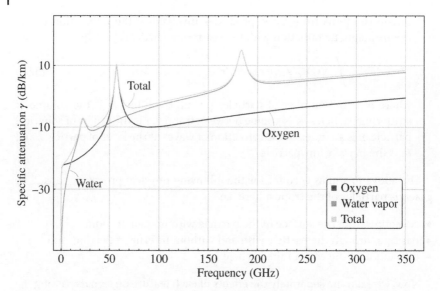

Figure 3.11 Specific attenuation of water vapor and oxygen (pressure = 1013 mbar, temperature = 15 °C, and water vapor content = 7.5 g/m³).

$\gamma_o(r)$ and $\gamma_w(r)$, at the ground level (where pressure is 1013 mbar and temperature is 15 °C), are given approximately by [29]

$$\gamma_o = \left(0.007\,19 + \frac{6.09}{f^2 + 0.227} + \frac{4.81}{(f - 57)^2 + 1.5} \right) 10^{-3} f^2 \tag{3.58}$$

$$\gamma_w = \left(0.05 + 0.0021\rho + \frac{10.6}{(f - 183.3)^2 + 9} + \frac{3.6}{(f - 22.2)^2 + 8.5} \right) 10^{-4} f^2 \tag{3.59}$$

where f is the frequency (GHz), and ρ is the water vapor density (g/m³).

Other temperatures are considered by correction factors of −1.0% per 1 °C from 15 °C for dry air and −0.6% per 1 °C from 15 °C for water vapor. The attenuation in the atmosphere over a path length L, for oxygen, L_o, and for water vapor, L_w, is given by [5, 9, 15, 29]

$$\begin{aligned} A_o &= \gamma_o L_o \\ A_w &= \gamma_w L_w \end{aligned} \tag{3.60}$$

The total atmospheric attenuation (in dB) for a particular path can then be found by integrating the total specific attenuation, as shown in Figure 3.12, over the total path length in the atmosphere [16] by assuming an exponential decrease in gas density with height [5, 9, 15, 29]:

$$L_a = \int_0^\pi \gamma_a(l)dl = \int_0^\pi [\gamma_o(l) + \gamma_w(l)]dl \text{ (dB)} \tag{3.61}$$

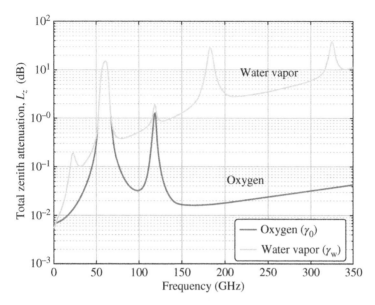

Figure 3.12 The total zenith attenuation vs. frequency for gaseous atmosphere with water vapor and dry air

The attenuation for an inclined path with an elevation angle $\theta > 10°$ can then be found from the zenith attenuation L_z as [5, 9, 15, 29]

$$L_a = L_z / \sin(\theta) \tag{3.62}$$

and was computed by Eqs. (3.3)–(3.8) following the parameters of the troposphere according to the standard ITU-676 [29].

3.4.2.3 Effects of rain
The attenuation of radio waves caused by rain increases with the number of raindrops along the radio path, the size of the drops, and the length of the path through which the rain passes, as shown in Figure 3.13, rearranged from [3].

There are several models for finding the attenuation caused by rain: empirical [23], semiempirical [24], and statistical–analytical models [5] (see also [17, 25, 26]). The Saunders's model which was embraced by ITU [24] does not depend on a particular place, is not frequency dependent, has a good processing time, and can be easily implemented.

In our work, we followed only the Saunders model. The Saunders model is applied when the density and shape of the raindrops are constant. According to [5], the received power P_r in a given antenna is found to drop exponentially with the radio path r through the rain and α as the reciprocal of distance required for the power decreases to e^{-1} of its initial value.

$$P_r(r) = P_r(0)e^{-\alpha r} \tag{3.63}$$

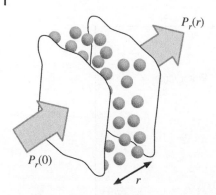

$P_r(r)$

Figure 3.13 Rain path attenuation, rearranged from [5]

$P_r(0)$

r

The value of α is given by the integral of one-dimensional (1D) distribution of the drops diameter D, denoted by $N(D)$, and $C(D)$ is the effective crosssection of the frequency-dependent signal power attenuation by rain drops.

$$\alpha = \int_0^\infty N(D) \cdot C(D) dD \tag{3.64}$$

In a real tropospheric situation, the drop diameter distribution $N(D)$ is not a constant value and can be found by the equation [5, 17, 23–26]

$$N(D) = N_0 \exp(-D/D_m) \tag{3.65}$$

where N_0 and D_m are parameters, and D_m depends on the rainfall rate remeasured above the ground surface in millimeters per hour. $N_0 = 8 \times 10^3$ m^{-2} mm^{-1}, and

$$D_m = 0.122 r^{0.21} \text{ mm} \tag{3.66}$$

As for $C(D)$, the attenuation crosssection from (3.64) can be found using Rayleigh approximation that is valid for lower frequencies [5, 17, 23–26], when the average drop size is smaller compared to the incident wavelength. In this case only absorption inside the drops occurs, and the Rayleigh approximation is valid giving a very simple expression for $C(D)$.

$$C[D] \propto \frac{D^3}{\lambda} \tag{3.67}$$

as was mentioned in [5, 24, 25], when $N(D)$ is not constant. As described earlier in (3.65), we take the value of the specific attenuation at a given point on the path, $\gamma(r)$, and integrate over the full path length R to find the total path loss:

$$L = \int_0^R \gamma(r) dr \tag{3.68}$$

while the total loss via the specific attenuation, as shown in [5, 17, 24], is defined by

$$\gamma = \frac{L}{r} = 4.34\alpha \tag{3.69}$$

Another way to describe the attenuation caused by rain is, when it increases more slowly with frequency approaching a constant value known as the *optical limit* [5]. Near this limit, scattering forms a significant part of attenuation that can be described using the Mie scattering theory that was described earlier [3, 4].

Expressing (3.69) in a logarithmic scale gives

$$L = 10 \log \left(\frac{P_T}{P_R} \right) \tag{3.70}$$

In practical situations, we can use an empirical model which implicitly combines all of these effects, where γ is assumed to depend only on distance r, whereas the rainfall rate (denoted by f) is measured on the ground in millimeters per hour. According to [5, 24, 25]

$$\gamma(f, R) = a(f) R^{b(f)} \tag{3.71}$$

The attenuation coefficients $a(f)$ and $b(f)$ can be found and calculated in [5, 24, 25]. The attenuation for a given path where the elevation angle θ is smaller than 90° makes it necessary to account for the variation in the rain in the horizontal direction. This allows us to focus on the finite size of rain clouds, called *rain cells*.

Also, rain varies in time over various parameters: seasonal, annual, and diurnal. It is important to realize that it is not the total amount of rain which falls during a given year that matters, but rather the period of time for which the rainfall rate exceeds a certain value. All of these temporal variations were estimated by use of (3.72) for rain attenuation, which does not exceed 0.01% of the time. Thus, according to [5, 17, 23–26]:

$$L_{0.01} = a(f) R_{0.01}^{b(f)} s_{0.01} r_R \tag{3.72}$$

where $s_{0.01}$ and rain path, r_R can be found in [5, 24, 25]. For time percentages other than 0.01%, the attenuation can be corrected by introducing a special relevant time percentage P, which is changed over the wide range from 0.001% to 1% [5, 24], that is

$$L(P) = L_{0.01} \cdot 0.12 P^{-(0.546 + 0.043 \log P)} \tag{3.73}$$

The precipitation of rain is defined by variations in both horizontal and vertical directions that make it very hard to describe the spatial distribution of rain. The correction factor we use in (3.72) is the effective path length (the product $s_{0.01} r_R$ in (3.72)).

The L_r, which is the length of the hypothetical path obtained from signal data, dividing the total attenuation by the specific attenuation exceeded for the same percentage of time [5]. It can be estimated, according to empirical model [5, 24, 25], as

$$L_r = \frac{L_s}{1 + 0.0286 L_h R^{0.15}} \tag{3.74}$$

Using Eqs. (3.71) and (3.74), we can now estimate that the transmission loss due to attenuation by rain is given by

$$A_r = \gamma_r L_r \tag{3.75}$$

3.4.2.4 Effects of clouds

As mentioned earlier, the dimension, shape, structure, and texture of clouds are influenced by air movements that change their formation and growth and the properties of the cloud particles. Sky cover is the observer view of the cover of the sky dome, whereas cloud cover can be used to describe areas that are smaller or larger than the floor space of the sky dome [16, 22]. There are several proposed models for the probability distribution of the sky cover [16, 22]. For our prediction of the cloud attenuation, we use the ITU-R model given in [22]. The specific attenuation due to a cloud can be determined by [22]

$$\gamma_c = K_1 M \left(\frac{dB}{km} \right) \tag{3.76}$$

where γ_c is the specific attenuation of the clouds, in dB/km; K_1 is the specific attenuation coefficient, in $(dB/km)/(g/m^3)$; and M is the liquid water density in g/m^3.

For small-sized cloud droplets, the *Rayleigh approximation* can be used for the calculation of specific attenuation [22]. This approximation is valid up to 100 GHz. A mathematical model based on Rayleigh scattering, which uses a double-Debye model for the dielectric permittivity $\varepsilon(f)$ of water, can be used to calculate the value of K_1:

$$K_1 = \frac{0.8917f}{\varepsilon''(1 + \eta^2)} \left(\frac{dB}{km} \right) \left(\frac{g}{m^3} \right)^{-1} \tag{3.77}$$

where f is the frequency in GHz, and η is defined as

$$\eta = \frac{2 + \varepsilon'}{\varepsilon''} \tag{3.78}$$

where ε' and ε'' are the real and imaginary components of the complex dielectric permittivity of water. For the calculation of the complex dielectric permittivity of water, we need to calculate the principal and secondary frequencies of the double Debye model for the dielectric permittivity of water [22]

$$\begin{aligned} f_p &= 20.09 - 142(\Phi - 1) + 294(\Phi - 1)^2 \\ f_s &= 590 - 1500(\Phi - 1) \end{aligned} \tag{3.79}$$

where $\Phi = 300/T$, and T is the temperature in kelvin. The complex dielectric permittivity of water is [22]

$$\varepsilon'(f) = \frac{\varepsilon_0 - \varepsilon_1}{1 + (f/f_p)^2} + \frac{\varepsilon_1 - \varepsilon_2}{1 + (f/f_s)^2} + \varepsilon_2 \tag{3.80}$$

$$\varepsilon''(f) = \frac{f(\varepsilon_0 - \varepsilon_1)}{f_p(1 + (f/f_p)^2)^2} + \frac{f(\varepsilon_1 - \varepsilon_2)}{f_s(1 + (f/f_s)^2)^2} + \varepsilon_2 \tag{3.81}$$

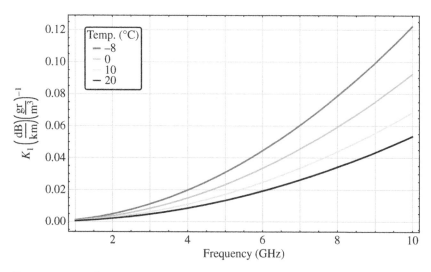

Figure 3.14 Specific attenuation of clouds as a function of frequency and temperature

and

$$\varepsilon_0 = 77.6 + 103.3(\Phi - 1), \quad \varepsilon_1 = 5.48, \quad \varepsilon_2 = 3.51$$

Figure 3.14, computed using the above formulas, shows the values of the specific attenuation K_1 at frequencies from 1 to 10 GHz and temperatures between −8 and +20 °C. As is clearly seen from the illustration in Figure 3.14, the specific attenuation of cloud parameter increases having polynomial form for frequencies of interests from 1 to 10 GHz, and this effect is observed for each temperature from −8 to +20 °C. With the increase in temperature, from −8 to +20 °C, the specific attenuation parameter K_1 of the cloud roughly doubles. Therefore, from summer time to winter time, the effect of clouds on signal attenuation during its passage inside land-atmospheric or atmospheric-land channels can be doubled.

Finally, we can determine the total cloud attenuation:

$$A = \frac{LK_1}{\sin(\theta)} \qquad (3.82)$$

where θ is the elevation angle ($5° \leq \theta \leq 90°$); K_1 is the specific attenuation coefficient as described earlier in Eq. (3.77), in (dB/km)/(g/m³); and L is the total columnar content of liquid water, in kg/m², or, equivalently, in mm of evaporated water.

Figure 3.15 shows the total cloud attenuation as a function of frequency, for elevation angles from 3° to 29°, according to the data taken from [11]. Our computations are based on liquid water content of 0.29 kg/m³.

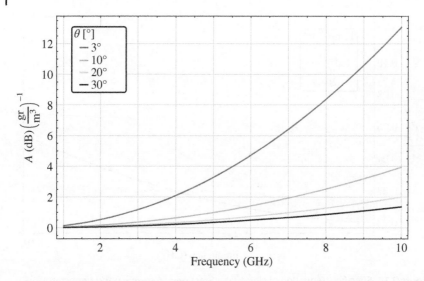

Figure 3.15 Total cloud attenuation as a function of frequency, for elevation angles from 3° to 30°, $T = 10\,°C$, and $L = 10$ km

3.4.2.5 Effects of turbulence

Atmospheric turbulence is a chaotic phenomenon created by the random temperature, wind magnitude variation, and direction variation [9, 15, 18, 19, 30–33]. This chaotic behavior, resulting in index-of-refraction fluctuations, causes Doppler shift and fast-fading phenomena. As is common for describing atmospheric turbulence, we use turbulence power spectra that are divided into three regions by two scale sizes. L_0, the outer scale of the turbulence, varies between 10 and 100 m and l_0, the inner scale, typically observed from 1 to 30 mm. The regions that are divided by those scales are called scintillations in the literature [9, 15, 18, 19, 30, 31].

Scintillation index The scintillation index (i.e. normalized variance of signal intensity fluctuations) describes fluctuations in optical power as measured by a point detector. The scintillation index is defined by [9, 15, 30, 31]

$$\sigma_I^2 = \frac{\langle I^2 \rangle - \langle I \rangle^2}{\langle I \rangle^2} = \frac{\langle I^2 \rangle}{\langle I \rangle^2} - 1 \tag{3.83}$$

The scintillation index relates to the Rytov variance (log-amplitude variance) σ_R^2 according to [4–7]. According to [24], the Rytov approximation starts from the premise that an air mass behaves as a fluid. Assuming that the refractive index structure parameter is constant, the basic Rytov approximation of relative variance is [9, 15, 18, 19, 30]

$$\sigma_R^2 = 1.23 \cdot C_n^2 \cdot k^{7/6} \cdot L^{11/6} \tag{3.84}$$

where L is the distance in km, C_n^2 is the refractive index structure parameter, and k is the wave number, $k = 2\pi/\lambda$. The dependence of the Rytov's refractive variance vs. the distance between the terminals of radio trace for various values of the refractive index structure parameter and for $f = 2.4, 3.3, 5.2$ GHz is presented in Figure 3.16a–c, respectively.

As can be seen from Figure 3.16a–c, much stronger turbulence (with increase of C_n^2) in the atmosphere leads to higher deviations of signal intensity variations – the effect increases nonlinearly with the increase in the range between the source and the detector. This increased effect is also clearly seen from the results of computations presented in Figure 3.17, where the same Rytov's scintillation index is presented as a function of C_n^2 for three frequencies, $f = 2.5, 3.3$, and 5.2 GHz, that are usually used in land-atmospheric communication networks (namely, in Wi-Fi wireless communication).

As can be seen from Figure 3.17, with the increase in frequency, the Rytov's scintillation index increases linearly as a function of C_n^2. Consequently, the fading effect becomes significant for signals passing through a turbulent atmospheric channel. Moreover, with an increase in frequency from 2.4 to 5.2 GHz, the scintillation index increases roughly twice, thus causing strong fading of signals passing through a turbulent tropospheric channel.

In [18, 32, 33] it was shown experimentally that the signal intensity scintillations, caused by quasi-local atmospheric turbulence, are distributed log-normally. In this case, it can be suggested that the fluctuations of the radio or optical signals are weak. The normalized standard deviation of this distribution is proportional to the Rytov's approximation described in (3.84) and can be written now via the structure parameter of turbulence permittivity C_ε^2 (instead of the structure parameter of refractivity C_n^2) as [9, 15, 18, 19, 30]

$$\sigma_I^2 = 0.12 \cdot C_\varepsilon^2 \cdot k^{7/6} \cdot L^{11/6} \tag{3.85}$$

We should note that C_ε^2 is the structure constant of the turbulence permittivity averaged over the path L, given in km, and k is the wave number mentioned above. In Figure 3.18, we present the computed index of signal intensity scintillations vs. the structure constant of the turbulence averaged over the path for different frequencies from 2.4 to 5.2 GHz.

It can be seen that approximately for $C_\varepsilon^2 = 10^{-10}$ the signal immediately starts to deteriorate, and as the frequency increases, the index of signal intensity scintillations becomes twice as strong. This result is very important for us because it helps us predict the fast fading of the signal within land aircraft radio communication links passing through the turbulent troposphere and operating at frequencies in the L/X-band ($f > 1 - 10$ GHz).

The fast fading of the signal at open paths is caused mainly by multipath propagation and turbulent fluctuations of the refractive index. The fluctuations of the signal intensity due to turbulence are distributed log-normally with the normalized standard deviation described by the Rytov variance described earlier in (3.86). For weak fluctuation with the Rytov method, the scintillation index

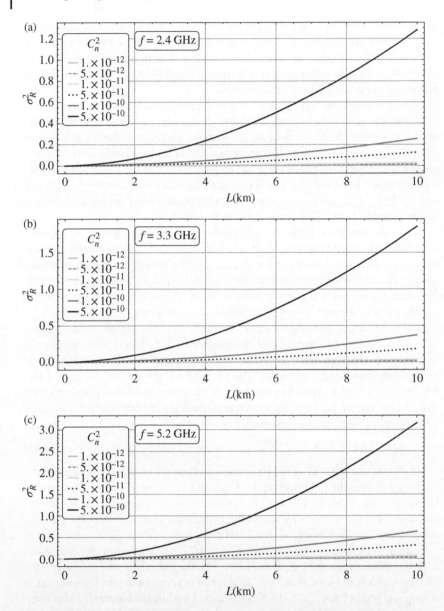

Figure 3.16 Rytov's scintillation index vs. distance between the terminals *L* for (a) *f* = 2.4 GHz, (b) *f* = 3.3 GHz, and (c) *f* = 5.2 GHz, for various refraction structure parameters of the turbulent atmosphere

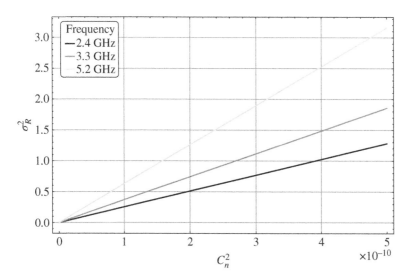

Figure 3.17 Rytov's scintillation index vs. C_n^2 for three frequencies, $f = 2.4, 3.3$, and 5.2 GHz for $L = 10$ km

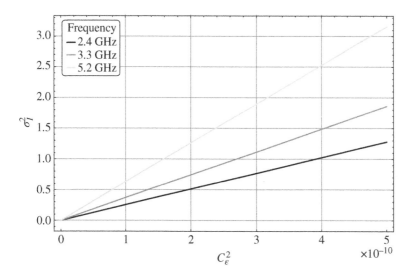

Figure 3.18 Index of signal intensity scintillations vs. the structure constant of the turbulence averaged over the path for frequencies from 1 to 50 GHz for $L = 10$ km

can be expressed in the following form [15, 30, 31]:

$$\sigma_I^2 = \exp(4\sigma_R^2) - 1 \tag{3.86}$$

The turbulence attenuation related to scintillation is equal to σ_R^2 (dB), and thus the relation for turbulence attenuation γ_R according to Rytov's theory of regular turbulence can be written as [15, 30, 31]

$$\gamma_R = 2 \cdot \sqrt{23.17 \cdot C_\varepsilon^2 \cdot k^{7/6} \cdot d^{11/6}} \qquad (3.87)$$

Relation of scintillation index and K-parameter of fast fading   Another way to calculate the attenuation due to fast-fading effects is to use the relations between K and the scintillation index σ_I. Usually in land wireless communication, instead of σ_I^2 the parameter of fading K is used [9, 15]. For Gaussian distribution described as zero-mean random process of turbulent structures evolution (usually observed experimentally in the irregular atmosphere), we could define the relation between the Ricean parameter of fading K, introduced above, and the scintillation index σ_I as

$$\langle \sigma_I^2 \rangle = \frac{\langle [I - \langle I \rangle]^2 \rangle}{\langle I \rangle^2} = \frac{I_{\text{inc}}^2}{I_{\text{co}}^2} \equiv K^{-2} \qquad (3.88)$$

where I_{co} and I_{inc} are the coherent and incoherent components of the total signal intensity. The results of computations according to (3.88) are shown in Figure 3.19.

The range $\langle \sigma_I^2 \rangle$ of the scintillation index variations, from 0.2 to 0.8, was obtained from numerous experiments, where relations between this parameter and the refractivity of the turbulence in the irregular atmosphere were taken into account (see [9, 15, 16]). Thus, from experiments described there, it was estimated that

$$C_n^2 \approx 10^{-15} \ (\text{m}^{-2/3}) \ \text{ and } \ C_n^2 \approx 10^{-13} \ (\text{m}^{-2/3}) \qquad (3.89)$$

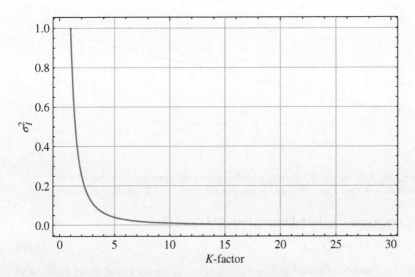

Figure 3.19 Scintillation index vs. Ricean fading parameter K

for a nocturnal and a diurnal atmosphere at the height around 1–2 km, respectively. As follows from (3.88), for $\langle \sigma_I^2 \rangle$ changing from 0.2 to 0.8, the fading parameter K changes from ~1.2–1.3 to ~3.5–3.8. This indicates the existence of direct visibility between both the terminals, the source and the detector, accompanied by the weak additional effects of multipath phenomena caused by multiple scattering of signals at the turbulent structures, formed in the disturbed atmospheric regions, observed experimentally [18, 32, 33]. In other words, a nonlinear relation between K and σ_I^2 states that when K is high, the scintillation index is low and vice versa; when σ_I^2 grows to its maximum value, parameter K reduces to its minimum value. When this occurs, we get the worst Rayleigh distribution and the biggest attenuation. The K-parameter can be used to determine the capacity, spectral efficiency, and BER of data stream sent via communication channel.

3.5 Link Budget Design

3.5.1 Path Loss in Free Space

The path loss for nonisotropic antennas (Tx and Rx) placed in the free space is given by [1–9]

$$L = 10 \log_{10} \left(\frac{J(r_2)}{J(r_1)} \right) = 10 \log_{10} \frac{A^2(r_2)}{A^2(r_1)} \tag{3.90}$$

where $J(r_1) = P_T$ and $J(r_2) = P_R$ are power (in watts) of the transmitter and the receiver which have directivity gain of G_T and G_R, respectively, and r is distance in km. From (3.90):

$$L = 10 \log_{10} \left(\frac{P_T}{P_R} \right) = 10 \log_{10} \left[\left(\frac{4\pi r}{\lambda} \right)^2 / (G_T G_R) \right] \tag{3.91}$$

Here, L_0 is the path loss of isotropic point source ($G_T = G_R = 1$) in free space and can be presented in (dB) as

$$L_0 = 10 \log_{10} \left(\frac{4\pi f r}{c} \right)^2 = 32.44 + 20 \log_{10} r + 20 \log_{10} f() \tag{3.92}$$

All the above formulas are related to the well-known Friis formula discussed in [16, 17]. Here, the frequency f is in MHz and the distance r in km.

3.5.2 Link Budget Design

Link budget is simply the maximum acceptable path loss and is usually split into two components, one of which is given by the distance-dependent path loss model plus a fade margin, which is included to allow the system some resilience against the practical effects of signal fading beyond the value predicted by the model.

$$\boxed{\begin{array}{c}\text{Maximum acceptable}\\\text{propagation loss}\end{array}} = \boxed{\begin{array}{c}\text{predicted}\\\text{loss}\end{array}} + \boxed{\begin{array}{c}\text{fade}\\\text{margin}\end{array}} \qquad (3.93)$$

For the tropospheric link, the total link budget is the sum of all the predicted losses such as the transmission line loss and the attenuation from the hydrometeors such as rain, clouds, and gaseous structures, mentioned above. As well as we account for fast fading caused by atmospheric turbulences and diffuse scattering and for slow fading caused by diffraction phenomenon from built-up terrain overlay profile. Finally, subtracting from this the gains of the antennas G_1 and G_2, the link budget can be given as [9, 15]

$$L_{\text{total}} = \overline{L} + L_{\text{SF}} + L_{\text{FF}} + \frac{S}{N_0} - G_T - G_R \qquad (3.94)$$

where

- L is the average path loss and is equal to the sum of free-space attenuation and NLOS condition attenuation, that is

$$\overline{L} = L_{\text{SF}} + L_{\text{NLOS}} \qquad (3.95)$$

- L_{NLOS} is the attenuation of hydrometeors such as rain, aerosols, and clouds;
- L_{FF} is the fast-fading loss due to effects of turbulence caused from scattering;
- L_{SF} is the slow-fading loss due to diffraction from built-up terrain profile;
- S/N is the signal-to-noise ratio accounting for the white Gaussian noise.

As examples on how to design link budget for atmosphere–land communication links, we design link budget for aircraft heights of 1, 5, 7, and 10 km, for three frequencies, $f = 2.4, 3.3,$ and 5.2 GHz, corresponding to that usually used in the fourth generation of the Wi-Fi Technology. In Tables 3.3–3.5, we conclude all the losses from hydrometeors, turbulences, and free space accounting for fast and slow fading. Data regarding environment was taken from middle latitude troposphere above three cities, small, medium, and large, in Israel. The gains of the transmission antenna and the aircraft antenna will be described as G_1 and G_2, and L is the horizontal distance. We explain the link budget for a distance $L = 10$ km to aircraft antenna (see Tables 3.3 and 3.4). The frequency range was taken corresponding to the actual Wi-Fi range bounded by 2.3–5.2 GHz.

As follows from Tables 3.3–3.5, the most impact in the total path loss, and therefore in link budget design, gives effect of the built-up terrain, the effect of which decreases with increase of the height of the moving vehicle. Thus, with increase of the height of the moving aircraft from 1 to 10 km, the effect of diffraction from buildings' overlay profile decreases approximately twice. In any case, the terrain factor plays much more important role in Wi-Fi communication networks design with respect to all tropospheric features – hydrometeors, gaseous structures, and turbulence.

Table 3.3 Link budget in the troposphere for $f = 2.4$ GHz and $L = 10$ km

Aircraft antennas height (km)	1	5	7	10
Tx antenna (dB)	G_1	G_1	G_1	G_1
Rx antenna (dB)	G_2	G_2	G_2	G_2
Free-space loss (dB)	120.0952	121.0211	121.7839	123.0623
Cloud loss (dB)	0.0175	0.0039	0.003	0.0025
Molecular loss (dB)	0.0727	0.0809	0.0883	0.1023
Rain loss (dB)	0.0238	0.0258	0.0277	0.0314
Fast fading (weak turbulence) (dB)	0.0014	0.00154	0.00167	0.00191
Fast fading(moderate turbulence) (dB)	0.31204	0.34407	0.37291	0.42678
Fast fading (strong turbulence) (dB)	0.44129	0.48659	0.52737	0.60355
Effect of the terrain profile				
Small city (dB) (1)	6.3	5.5	4.6	3.9
Medium city (dB) (2)	9.5	7.7	5.8	4.7
Large city (dB) (3)	14.2	12.9	10.4	8.8
Total path loss (dB) (1)	127.2639	127.4639	127.4049	128.1307
(2)	130.4639	129.6639	128.6049	128.9307
(3)	135.1639	134.8639	133.2049	133.0307

Table 3.4 Link budget in the troposphere for $f = 3.3$ GHz and $L = 10$ km

Aircraft antennas height (km)	1	5	7	10
Tx antenna (dB)	G_1	G_1	G_1	G_1
Rx antenna (dB)	G_2	G_2	G_2	G_2
Free-space loss (dB)	120.0952	121.0211	121.7839	123.0623
Cloud loss (dB)	0.033	0.0073	0.0057	0.0046
Molecular loss (dB)	0.0781	0.0869	0.0949	0.1099
Rain loss (dB)	0.0478	0.0512	0.0548	0.0621
Fast fading (weak turbulence) (dB)	0.00168	0.00185	0.00201	0.0023
Fast fading (moderate turbulence) (dB)	0.37574	0.41431	0.44904	0.5139
Fast fading (strong turbulence) (dB)	0.53137	0.58592	0.63503	0.72676
Effect of the terrain profile				
Small city (dB) (1)	6.3	5.5	4.6	3.9
Medium city (dB) (2)	9.5	7.7	5.8	4.7
Large city (dB) (3)	14.2	12.9	10.4	8.8
Total path loss (dB) (1)	130.229	130.4347	130.3914	131.148
(2)	133.429	132.6347	131.5914	131.948
(3)	138.129	137.8347	136.1914	136.048

Table 3.5 Link budget in the troposphere for f = 5.2 GHz and L = 10 km

Aircraft antennas height (km)	1	5	7	10
Tx antenna (dB)	G_1	G_1	G_1	G_1
Rx antenna (dB)	G_2	G_2	G_2	G_2
Free-space loss (dB)	120.0952	121.0211	121.7839	123.0623
Cloud loss (dB)	0.082	0.0182	0.0142	0.0115
Molecular loss (dB)	0.0908	0.101	0.1103	0.1278
Rain loss (dB)	0.2907	0.3119	0.3323	0.3725
Fast fading (weak turbulence) (dB)	0.00219	0.00242	0.00262	0.003
Fast fading (moderate turbulence) (dB)	0.48988	0.54016	0.58544	0.67001
Fast fading (strong turbulence) (dB)	0.69279	0.7639	0.82794	0.94753
Effect of the terrain profile				
Small city (dB) (1)	6.3	5.5	4.6	3.9
Medium city (dB) (2)	9.5	7.7	5.8	4.7
Large city (dB) (3)	14.2	12.9	10.4	8.8
Total path loss (dB) (1)	134.7595	134.9746	134.9725	135.8105
(2)	137.9595	137.6347	136.1725	136.6105
(3)	143.129	142.3746	140.7725	140.7105

References

1 Rappaport, T.S. (1996). *Wireless Communications: Principles and Practice*, 2e in 2001. Noboken, NJ: Prentice Hall PTR.

2 Ponomarev, G.A., Kulikov, A.N., and Telpukhovsky, E.D. (1991). *Propagation of Ultra - Short Waves in Urban Environments*. Tomsk, Rasko: USSR.

3 Blaunstein, N. (2004). Wireless communication systems. In: *Handbook of Engineering Electromagnetics*, Chapter 12 (ed. R. Bansal). New York, NY: Marcel Dekker. 417–482.

4 Bertoni, H.L. (2000). *Radio Propagation in Modern Wireless Systems*. Upper Saddle River, NJ: Prentice Hall PTR.

5 Saunders, S.R. (2001). *Antennas and Propagation for Wireless Communication Systems*. Chichester: Wiley.

6 Goodman, D.J. (1997). *Wireless Personal Communication Systems*. Reading, MA: Addison-Wesley.

7 Schiller, J. (2003). *Mobile Communications, Addison-Wesley Wireless Communications Series*, 2e. Reading, MA: Addison-Wesley.

8 Molisch, A.F. (2007). *Wireless Communications*. Chichester: Wiley.

9 Blaunstein, N. and Christodoulou, C. (2007). *Radio Propagation and Adaptive Antennas for Wireless Communication Links*, 1e. New Jersey: Wiley.

10 Blaunstein, N., Katz, D., Censor, D. et al. (2002). Prediction of loss characteristics in built-up areas with various buildings' overlay profiles. *J. Anten. Propag. Mag.* 44(1): 181–192.

11 Blaunstein, N. and Levin, M. (1996). VHF/UHF wave attenuation in a city with regularly spaced buildings. *Radio Sci.* 31 (2): 313–323.

12 Blaunstein, N. (1999). Prediction of cellular characteristics for various urban environments. *J. Anten. Propag. Mag.* 41 (6): 135–145.

13 Blaunstein, N. (1998). Average field attenuation in the non-regular impedance street waveguide. *IEEE Trans. Anten. Propag.* 46 (12): 1782–1789.

14 Yarkoni, N., Blaunstein, N., and Katz, D. (2007). Link budget and radio coverage design for various multipath urban communication links. *Radio Sci.* 42 (2): 412–427.

15 Blaunstein, N. and Christodoulou, C. (2014). *Radio Propagation and Adaptive Antennas for Wireless Communication Networks - Terrestrial, Atmospheric and Ionospheric*, 2e. New Jersey: Wiley.

16 Chou, M.D. (1998). Parameterizations for cloud overlapping and shortwave single scattering properties for use in general circulation and cloud ensemble models. *J. Clim.* 11: 202–214.

17 Crane, R.K. (1980). Prediction of attenuation by rain. *IEEE Trans. Commun.* 28: 1717–1733.

18 Blaunstein, N., Arnon, Sh., Zilberman, A., and Kopeika, N. (2010). *Applied Aspects of Optical Communication and LIDAR*. New York: CRC Press.

19 Blaunstein, N. and Kopeika, N. (eds.) (2018). *Optical Waves and Laser Beams in the Irregular Atmosphere*. CRC Press: Boca Raton, FL.

20 d'Almeida, G.A., Koepke, P., and Shettle, E.P. (1991). *Atmospheric Aerosols, Global Climatology and Radiative Characteristics*. Hampton, VA: Deepak Publishing.

21 Seinfeld, J.H. (1986). *Atmospheric Chemistry and Physics of Air Pollution*. New York: Wiley.

22 ITU-R Recommendation International Telecommunication Union (1992). Attenuation due to clouds and fog, pp. 840–842.

23 Lin, D.P. and Chen, H.Y. (2002). An empirical formula for the prediction of rain attenuation in frequency range 0.6-100 GHz. *IEEE Trans. Anten. Propag.* 50: 545–551.

24 International Telecommunication Union, ITU-R Recommendation, P.838 (1992). Specific attenuation model for rain for use in prediction methods, Geneva.

25 ITU-R Recommendation International Telecommunication Union, P.838 (1992). Specific attenuation model for rain for use in prediction methods.

26 Kooi, P.-S., Leong, M.-S, Li, L.-W et al. (1995). Microwave attenuation by realistically distorted raindrops: Part II - predictions. *IEEE Trans. Anten. Propag.* 43: 821–828.

27 Jaenicke, R. (1988). Aerosol physics and chemistry. In: *Physical Chemical Properties of the Air, Geophysics and Space Research*, vol. 4(b) (ed. G. Fisher). Berlin: Springer-Verlag. 153–187

28 Rosen, J.M. and Hofmann, D.J. (1986). Optical modeling of stratospheric aerosols: present status. *Appl. Opt.* 25(3): 410–419.

29 ITU-R International Telecommunication Union, ITU-R Recommendation (1997). Attenuation by atmospheric gases, pp. 676–683.

30 Ishimaru, A. (1978). *Wave Propagation and Scattering in Random Media.* New York: Academic Press.

31 Andrews, L.C. and Phillips, R.L. (2005). *Laser Beam Propagation through Random Media*, 2e. Bellingham, WA: SPIE Press.

32 Bendersky, S., Kopeika, N., and Blaustein, N. (2004). Prediction and modeling of line-of-sight bending near ground level for long atmospheric paths. *Proceedings of SPIE International Conference*, San Diego, CA (3–8 August 2004), pp. 512–522.

33 Bendersky, S., Kopeika, N., and Blaustein, N. (2004). Atmospheric optical turbulence over land in middle-east coastal environments: prediction, modeling and measurements. *J. Appl. Opt.* 43: 4070–4079.

4

Polarization Diversity Analysis for Networks Beyond 4G

Depolarization of electromagnetic (EM) waves in terrestrial communication links has been poorly researched even for wireless networks beyond 3G. The main source of knowledge to this field comes from Stokes parameters and various field test results [1–4]. Below, we present an innovative analytical approach based on the multiparametric stochastic model, as a combination of physical vision on radio propagation above the built-up terrain and stochastic description of main features of the terrain, discussed in Chapters 4 and 5, to this topic, providing a sufficient way of predicting the polarization of radio waves in such communication links. This can be used to reduce polarization losses in receiving antennas, and hence, more effective data transmission can be achieved. This approach allows to predict the radio wave total intensity distribution caused by built-up obstructions based on probability functions of the spatial polarization ellipse.

4.1 Depolarization Phenomena in Terrain Channels

Roughly speaking, depolarization is the deformation of the polarization ellipse due to diffraction back off from multiple obstacles, as illustrated in Figure 4.1, or propagation in space where the surrounding media affect the polarization.

In a communication link with obstacles, the absorbed signal might be polarized utterly differently, compared to its original form. Four main characteristic variations of the diffracted wave may be considered: the semiaxes lengths, the angle of rotation, inclination, or azimuth angle. In fact, one can predict, to some point of precision, the value of these parameters.

Due to polarization losses, the most important parameter is the angle of rotation of the polarization ellipse. Figure 4.2 shows that polarization loss is the mismatch between the orientations of the arriving wave and the antenna [3, 4].

This mismatch angle, the ratio between the polarization ellipse's axes and the inclination angle, can be calculated using the formulas which will be presented later in this chapter. Since no information on the orientation of a receiver

Advanced Technologies and Wireless Networks Beyond 4G, First Edition.
Nathan Blaunstein and Yehuda Ben-Shimol.
© 2021 John Wiley & Sons, Inc. Published 2021 by John Wiley & Sons, Inc.

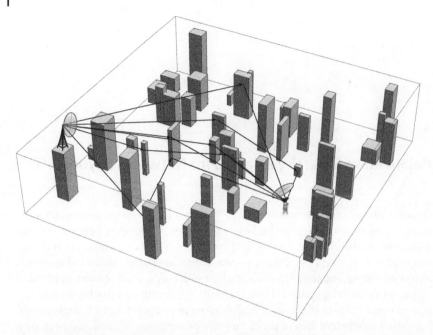

Figure 4.1 Depolarization in built-up areas – a circularly polarized electromagnetic wave depolarized after hitting multiple objects between the transmitting and receiving antennas

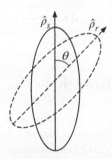

Figure 4.2 Polarization unit vectors and the angle of mismatch

antenna is provided, we assume that the antenna is perpendicular to the ground and is the angle of mismatch.

4.2   Model by Stocks Parameters

Following[1–3], one can write an expression for the electric part of the electromagnetic wave in space, using the quadrature components of the field and relations between them. Consider a reference orthogonal system of coordinates with unit vectors . The electrical field may be represented by the cosine and sine components and by [1–3]

$$E(t) = \mathbf{S} \cos \omega t + \mathbf{C} \sin \omega t \tag{4.1}$$

where

$$\mathbf{S} = -\sum_{i=1}^{3} \mathbf{E}_i \sin(\varphi_i) \tag{4.2}$$

$$\mathbf{C} = \sum_{i=1}^{3} \mathbf{E}_i \cos(\varphi_i) \tag{4.3}$$

The normal vector in the direction of propagation can be written as

$$\mathbf{N} = \mathbf{S} \times \mathbf{C}$$

with magnitude

$$|\mathbf{N}| = |\mathbf{C}|^2 |\mathbf{S}|^2 - (\mathbf{C} \cdot \mathbf{S})^2 \tag{4.4}$$

Considering an elliptical polarization and using the Pauli spin matrices, noted in [2, 3] and expressions (4.2) and (4.3), the Stokes parameters can be rewritten as

$$I = \sum_{i=1}^{3} (C_i^2 + S_i^2) \tag{4.5}$$

$$Q = I - 2I_3 / \sin^2(\theta) \tag{4.6}$$

$$U = 2 \, (N_1 \cos(\varphi) + N_2 \sin(\varphi)) / \sin(\varphi) \tag{4.7}$$

$$V = 2(\mathbf{C} \times \mathbf{S}) = 2\mathbf{N} \tag{4.8}$$

where $I_i = C_i^2 + S_i^2$, $i = 1, 2, 3$ and θ and φ are the azimuthal and polar angles of a spherical coordinate system. Using expressions (4.5)–(4.8), a joint probability density function (PDF) of all six quadrature components of the polarization ellipse is presented assuming that all six parameters are normally distributed independent random variables [1, 2]. Therefore, the PDF is given by

$$\varphi(C, S) = \prod_{i=1}^{3} \frac{1}{2\pi\sigma_i} \exp\left(-\frac{C_i^2 + S_i^2}{2\sigma_i^2}\right) \tag{4.9}$$

The polar and azimuthal angles $\varpi \in [0, 2\pi]$ and $\theta \in [0, \pi]$ are fixing the position of the vector $\mathbf{N}$ that is normal to the polarization ellipse. The projections of $\mathbf{N}$ on the coordinate systems $\hat{\mathbf{u}}_1$, $\hat{\mathbf{u}}_2$, and $\hat{\mathbf{u}}_3$, are N_1, N_2, and N_3, respectively, and relate to θ and φ by

$$\theta = \tan^{-1}(N_2 / N_1) \tag{4.10}$$

$$\varphi = \tan^{-1}(N_3 / N_1) \tag{4.11}$$

The relevant vectors and directional parameters are shown in Figure 4.3.

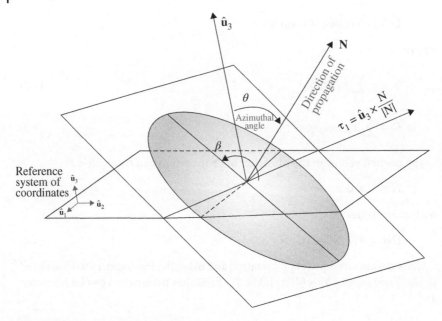

Figure 4.3 Geometry of the problem

The parameter R is the ratio of the minor to the major axes of the polarization ellipse

$$R = 2N/(I + \sqrt{I^2 - 4N^2}) \tag{4.12}$$

However, the following ratio is more commonly used:

$$p = \frac{1 - R^2}{1 + R^2} \tag{4.13}$$

Finally, (4.9) can be rewritten as a function of the parameters which fully describe the spatial polarization ellipse

$$\omega(I, p, \varphi, \theta, \beta) = \frac{I^2 p \sin(\theta)}{(2\pi)^3} \prod_{i=1}^{3} \frac{1}{2\pi\sigma_i} \exp\left(-\frac{I_i^2}{2\sigma_i^2}\right) \tag{4.14}$$

β is the angle of inclination, that is, the angle between the large axis of the polarization ellipse and $\hat{\mathbf{u}}_3 \times \mathbf{N}$ (e.g. assuming that the ground plane is described by $\hat{\mathbf{u}}_1, \hat{\mathbf{u}}_2$ and the orientation of the antenna is $\hat{\mathbf{u}}_3$). Expression (4.14) is less applicable to the investigation of communication links, in the sense that it does not account for any channel characteristics. However, $\omega(I, p, \varphi, \theta, \beta)$ depends on the signal's intensity, which, in turn, can be described by the channel's characteristics.

4.3 The Multiparametric Stochastic Model Application for Polarization Parameters Prediction

In Chapter 3, the wave's average total intensity as a function of a channel's characteristics, using its coherent and incoherent parts following [1–3], was represented briefly. For the reader's convenience, we repeat it once more. The coherent part is a deterministic value which is less affected by the communication link compared to the incoherent part.

$$\langle I_{\text{total}} \rangle = \langle I_{\text{co}} \rangle + \langle I_{\text{inc}} \rangle \tag{4.15}$$

where

$$\langle I_{\text{co}} \rangle = \exp\left(-\gamma_0 d \frac{\overline{h} - z_1}{z_2 - z_1} \right) \cdot \left(\frac{\sin(kz_1 z_2/d)}{2\pi d} \right)^2 \tag{4.16}$$

$$\langle I_{\text{inc}} \rangle = \frac{\Gamma \lambda^2 \ell_h \ell_v (z_2 - \overline{h})}{8\pi[\lambda^2 + (2\pi \ell_h \overline{L} \gamma_0)^2][\lambda^2 + (2\pi \ell_v \gamma_0(\overline{h} - z_1))^2]d^3} \tag{4.17}$$

See Table 4.1 for a more complete description of the channel parameters. The incoherent part can be decomposed to describe the energy variance in two orthogonal planes: parallel and perpendicular to the wave's plane [1, 2].

$$\langle I_{\text{inc}} \rangle = \sigma_\perp^2 \cdot \sigma_\parallel^2 \tag{4.18}$$

This is shown in Figure 4.4.

Table 4.1 Parameter values for mixed-residential and suburban zones

Parameter	Meaning	Mixed residential	Suburban
z_1	Receiver height	2 m	2 m
z_2	Transmitter height	35 m	45 m
d	Distance	0.5–5 km	0.2–2 km
h_{min}	Minimum height	5 m	15 m
h_{max}	Maximum height	10 m	40 m
$\overline{L}$	Average length	20–40 m	50–60 m
ℓ_h	Horizontal coefficient	1 m	2 m
ℓ_v	Vertical coefficient	1 m	2 m
v	Aerial buildings' density	30–40 km^{-2}	50–60 km^{-2}
γ_0	Obstacles line density	0.4, 0.7, 1	1.6, 2, 2.3
Γ	Reflection coefficient	0.7 (stone)	0.2 (glass)
n	Obstacle height index	0.05, 0.1	1, 10, 20

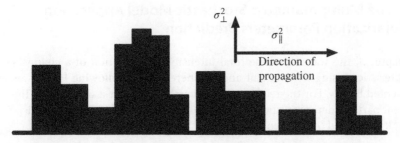

Figure 4.4 Terrestrial link profile and the components of the energy variances

Since $\sigma_\perp$, $\sigma_\parallel$ are defined as orthogonal, their expressions are obtained by the attributing horizontal and vertical components of $\langle I_{\text{inc}} \rangle$, respectively:

$$\sigma_\perp^2 = \frac{\Gamma}{8\pi} \cdot \frac{\lambda \ell_h}{\lambda^2 + (2\pi \ell_h \gamma_0 \overline{L})^2} \cdot \frac{1}{d} \tag{4.19}$$

$$\sigma_\parallel^2 = \frac{\lambda \ell_v}{\lambda^2 + (2\pi \ell_v \gamma_0 (h - z_1))^2} \frac{z_2 - \overline{h}}{d^2} \tag{4.20}$$

The function $P_h(z)$ is the complementary cumulative distribution function (CCDF) determining the probability of a building being lower than a given height h, i.e. $z < h$, where z denotes the antenna's height. We present again the buildings' height profile inside the city layer according to [3, 4]:

$$P_h(z) = H(h_1 - z) + H(z - h_1)H(h_2 - z)\left[\frac{h_2 - z}{h_2 - h_1}\right]^n \tag{4.21}$$

where $n > 0, 0 < z < h_2, h_1$ and h_2 are the minimum and maximum of the building heights, respectively, and n is an indicator of the ratio between the numbers of low and high buildings. The function $H(\cdot)$ is the Heaviside unit step function. Using (4.21), a function representing the build-up profile for the area between two antennas can be represented according to [3, 4] as

$$F(z_1, z_2) = \int_{z_1}^{z_2} P_h(z)dz \tag{4.22}$$

The "shadow function" (4.22) represents the probability of a signal to reach a client that is present in the shadow of obstacles in a terrain. Using (4.22), expressions (4.19) and (4.20) are rewritten to achieve the average signal intensity in suburban areas for a single ray:

$$\langle I_{\text{inc}} \rangle = \frac{\Gamma}{8\pi} \cdot \frac{\lambda \ell_v \sqrt{\frac{\lambda d}{4\pi^3} + (z_2 - \overline{h})^2}}{[\lambda^2 + (2\pi \ell_v \gamma_0 F(z_1, z_2))^2]} \cdot \frac{1}{d^3} \tag{4.23}$$

and for two-ray scatterings:

$$\langle I_{inc} \rangle = \frac{\Gamma^2}{24\pi^2} \cdot \frac{\lambda^3 \ell_v^2 \left(\frac{\lambda d}{4\pi^3} + (z_2 - \overline{h})^2 \right)}{[\lambda^2 + (2\pi \ell_v \gamma_0 F(z_1, z_2))^2]^3} \cdot \frac{1}{d^3} \tag{4.24}$$

Every additional ray at the multipath is further attenuated in around 15 dB; thus, we find it sufficient to analyze up to two rays. For the simplified case of areas, where most of the buildings' heights are distributed uniformly (i.e. $n \approx 1$):

$$\langle I_{inc} \rangle = \frac{\Gamma}{8\pi} \cdot \frac{\lambda \ell_h}{\lambda^2 + (2\pi \ell_h \gamma_0 \overline{L})^2} \cdot \frac{\lambda \ell_v \sqrt{\frac{\lambda d}{4\pi^3} + (z_2 - \overline{h})^2}}{\lambda^2 + (2\pi \ell_v \gamma_0 (\overline{h} - z_1))^2} \cdot \frac{1}{d^3} \tag{4.25}$$

Now we redo the process above, dividing expressions (4.23)–(4.25) into the energy variances. From (4.23), we get, according to [3, 4],

$$\sigma_{\parallel}^2 = \frac{\lambda \ell_v \sqrt{\frac{\lambda d}{4\pi^3} + (z_2 - \overline{h})^2}}{\lambda^2 + [2\pi \ell_v \gamma_0 F(z_1, z_2)]^2} \cdot \frac{1}{d^2} \tag{4.26}$$

$$\sigma_{\perp}^2 = \frac{\Gamma}{8\pi d} \tag{4.27}$$

And for two rays, (4.24) leads to

$$\sigma_{\parallel}^2 = \frac{\Gamma}{3\pi} \frac{\lambda^3 \ell_v^2 \left(\frac{\lambda d}{4\pi^3} + (z_2 - \overline{h})^2 \right)}{[\lambda^2 + (2\pi \ell_v \gamma_0 F(z_1, z_2))^2]^3} \frac{1}{d^2} \tag{4.28}$$

$$\sigma_{\perp}^2 = \frac{\Gamma}{8\pi d} \tag{4.29}$$

For terrains with $n \approx 1$, (4.25) is represented by

$$\sigma_{\parallel}^2 = \frac{\lambda \ell_v \sqrt{\frac{\lambda d}{4\pi^3} + (z_2 - \overline{h})^2}}{\lambda^2 + (2\pi \ell_v \gamma_0 (\overline{h} - z_1))^2} \cdot \frac{1}{d^2} \tag{4.30}$$

$$\sigma_{\perp}^2 = \frac{\Gamma}{8\pi} \frac{\lambda \ell_h}{\lambda^2 + (2\pi \ell_h \gamma_0 \overline{L})^2} \cdot \frac{1}{d} \tag{4.31}$$

The effect of the terrain on the EM wave may be better understood by the ratio $\Delta = (\sigma_{\|}/\sigma_{\perp})^2$.

$$
\Delta = \left(\frac{\sigma_{\|}}{\sigma_{\perp}}\right)^2 = \begin{cases} \dfrac{8\pi\ell_v\sqrt{\dfrac{\lambda d}{4\pi^3} + (z_2 - \overline{h})^2}}{\Gamma[\lambda^2 + (2\pi\ell_v\gamma_0 F(z_1, z_2))^2]d} & \text{single ray} \\[2em] \dfrac{8\lambda^3\ell_v^2\left[\dfrac{\lambda d}{4\pi^3} + (z_2 - \overline{h})^2\right]}{3[\lambda^2 + (2\pi\ell_v\gamma_0 F(z_1, z_2))^2]^3 d} & \text{two rays} \\[2em] \dfrac{8\pi\ell_v}{\Gamma\ell_h}\dfrac{[\lambda^2 + (2\pi\ell_h\gamma_0\overline{L})^2]\sqrt{\dfrac{\lambda d}{4\pi^3} + (z_2 - \overline{h})^2}}{[\lambda^2 + (2\pi\ell_v\gamma_0(\overline{h} - z_1))^2]d} & n \approx 1 \end{cases}
$$

$$(4.32)$$

Finally, from (4.17) to (4.19), we present the modified form of (4.14)

$$
w(I, p, \varphi, \theta, \beta) = \frac{I^2 p \sin(\theta)}{8(2\pi)^2\sigma_{\perp}^4\sigma_{\|}^3} \cdot \exp\left[-\frac{I}{2}\left(\frac{1 - \xi\sin^2\theta}{\sigma_{\perp}^2} - \frac{\xi\sin^2\theta}{\sigma_{\|}^2}\right)\right]
$$

$$(4.33)$$

Integrating (4.33) over each parameter yields the one-dimensional probability functions for the parameters that describe the polarization ellipse in space [2, 3]. Then, the marginal distributions $w(\theta)$, $w(\beta)$, and $w(R)$ are

$$
w(\theta) = \Delta\frac{p\sin\theta}{2[\cos^2\theta - \Delta^2\sin^2\theta]^{3/2}}
$$

$$(4.34)$$

$$
w(\beta) = \frac{1}{\pi}\Delta\frac{(1 + \eta)^2(1 + \eta_0)}{(\eta - \eta_0)^2} \cdot
$$
$$
\left[\frac{\eta - 3\eta_0}{2}\sqrt{\eta_0}\tan^{-1}(\sqrt{\eta}) + \eta_0^{3/2}\tan^{-1}(\sqrt{\eta_0}) + \frac{\eta}{2}\frac{\eta - \eta_0}{1 + \eta}\right] \quad (4.35)
$$

$$
w(R) = \frac{16C(1 - R^2)}{(1 + R^2)^3}\left[\frac{2 + \eta_0^3}{1 - (p\eta_0)^2}\right]
$$

$$(4.36)$$

where $\eta_0 = (\sigma_{\|}^2 - \sigma_{\perp}^2)/(\sigma_{\|}^2 + \sigma_{\perp}^2) = (\Delta - 1)/(\Delta + 1)$, $\eta = (\eta_0(1 - \cos(2\beta)))/(1 + \eta_0\cos(2\beta))$, and $C = 2\eta_0^2(1 - \eta_0^2)/(\eta_0^4 - 2(1 - \eta_0^2)\ln(1 - \eta_0^2))$. Expression (4.35) is the one considered most relevant for predicting depolarization losses.

4.4 Numerical Analysis of Probability Functions for Parameters of the Spatial Polarization Ellipse

Expressions (4.32), (4.34)–(4.36) are the most relevant for simulation; however, in this work we focus at investigating expression (4.35). First, the relevant values for the energy variances $\sigma_{\|}^2$, $\sigma_{\perp}^2$ need to be calculated and then use them in (4.35). Since the number of different parameters and their combinations is huge, for the present work we use the parameters given in Table 4.1, with frequencies $f = 0.9, 1.8, 2.4$, and 5.2 GHz. For the mixed-residential areas, we use the two-ray submodel described by (4.32), as it is more realistic than the single-ray model.

4.4.1 Mixed-residential Areas

Mixed-residential areas characterized by a very sparse density of obstacles and high obstacles are rare. Figure 4.5 shows the ratio Δ as a function of distance between antennas and the carrier frequency, for $\gamma_0 = 0.4$ km^{-1} and average building height $\bar{h} = 10$ m.

Figure 4.5 shows that Δ decreases as a function of distance. Moreover, for a given distance, Δ increases with frequency. Near the transmitting antenna (i.e. small d) and for high frequencies $\Delta \gg 1$, that is, most of the energy of the incoherent part is gathered in a plane perpendicular to the plane of the ellipse, while for large distances from the antenna, this behavior is reversed. Next, using (4.35), we calculate the PDF $\omega(\beta)$ for the angle of inclination for four common carrier frequency values 900 MHz, 1.8 GHz, 2.4 GHz, and 5.2 GHz. The lack

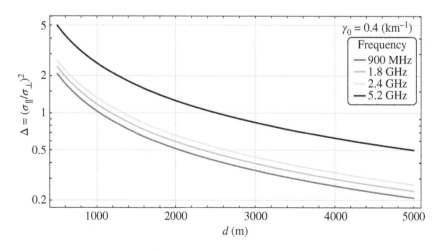

Figure 4.5 The ratio $\Delta = \left(\frac{\sigma_{\|}}{\sigma_{\perp}}\right)^2$ as a function of distance and carrier frequency in mixed-residential areas

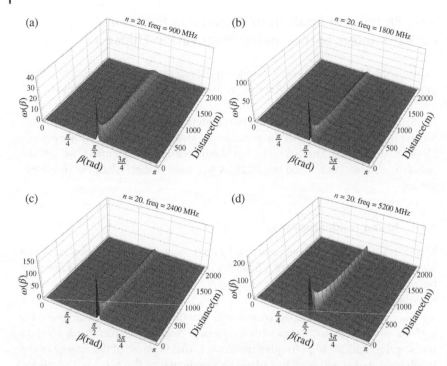

Figure 4.6 The PDF $\omega(\beta)$ in mixed-residential areas. (a) $n = 20$ (many more low buildings), $\gamma_0 = 0.4$ km^{-1}, $f = 900$ MHz. (b) $n = 20$ (many more low buildings), $\gamma_0 = 0.4$ km^{-1}, $f = 1.8$ GHz. (c) $n = 20$ (many more low buildings), $\gamma_0 = 0.4$ km^{-1}, $f = 2.4$ GHz. (d) $n = 20$ (many more low buildings), $\gamma_0 = 0.4$ km^{-1}, $f = 5.2$ GHz

of tall buildings is represented by the value $n = 20$ and the sparsity of buildings (houses) by the small line density of buildings by $\gamma_0 = 0.4$ km^{-1}. The results of numerical simulations are presented in Figure 4.6.

Note that due to the symmetry of the ellipse, a full rotation of the ellipse happens already for $\beta = 180°$. Also, due to the symmetry of $\omega(\beta)$, the expected value of β is always equal to $\pi/2$, independent of the frequency or other parameters of the area. However, the variance of the PDF is both frequency and distance dependent. We mention again that (4.35) is a PDF as a parameter of β for a fixed distance d. For small distances from the transmitting antenna, the polarization ellipse for the incoherent part is most likely inclined around $\beta = 90°$. As the distance increases, the standard deviation of increases and that probability of an inclination angle far from increases. This tendency is also frequency dependent, as shown in Figure 4.7. Far from the antenna and for high frequencies (i.e. 5.2 GHz, continuous curve) $\omega(\beta)$ keeps the bell shape, while for low carrier frequency (900 MHz, continues curve) the probability of inclination angles near $0°$ increases. This implies that, for low carrier frequencies and far from the transmitting antennas, a horizontally polarized receiving antenna will be

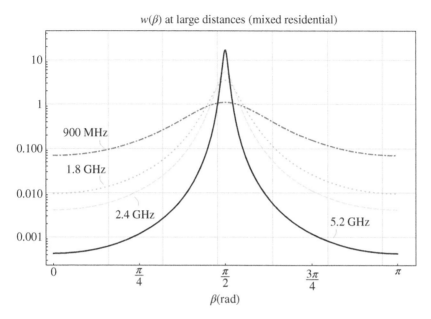

Figure 4.7 The PDF $\omega(\beta)$ in mixed-residential areas far from the transmitter

affected more from the incoherent part of the received wave. As the frequency goes higher, a vertically polarized receiving antenna will be more affected by the incoherent part.

In addition, with the increase in distance, it is more difficult to predict the probable value of the inclination angle as the PDF becomes "flatter." However, values of 90° are more probably closer to the transmitting antenna. Also, both increase in building density and carrier frequency lower the probability for 0° or 180° rotations.

4.4.2 Suburban and Urban Areas

In suburban or urban areas(as fully built-up areas), the transmitting antenna is usually higher, for example, $z_2 = 45$ m, high buildings are more common than low ones (i.e. small n), and the buildings' density is higher. Figure 4.8 shows a much higher ratio Δ than in mixed-residential areas, where the LOS component is much higher (low buildings) .

Figure 4.9 shows the similarity (in shape) of $\omega(\beta)$ to the mixed-residential case, but it is clear that the central peak of the PDF decays more slowly, as distance increases, implying more ray-like energy distribution, or, in other words most of the energy is contained in the $\sigma_\perp^2$ part. Likewise, in the case of mostly high buildings (i.e. small n), we maintain the same behavior. But as the obstacles get lower, that is, in the case of high n, we see that the ratio Δ grows.

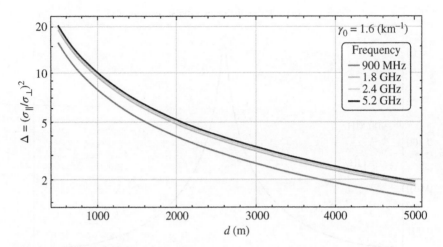

Figure 4.8 The ratio Δ as a function of distance and carrier frequency in suburban areas

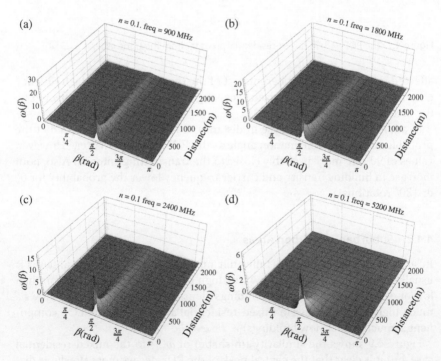

Figure 4.9 The PDF $\omega(\beta)$ in suburban areas. (a) $n = 0.1$ (more high buildings), $\gamma_0 = 1.6$ km^{-1}, $f = 900$ MHz. (b) $n = 0.1$ (more high buildings), $\gamma_0 = 1.6$ km^{-1}, $f = 1.8$ GHz. (c) $n = 0.1$ (more high buildings), $\gamma_0 = 1.6$ km^{-1}, $f = 2.4$ GHz. (d) $n = 0.1$ (more high buildings), $\gamma_0 = 1.6$ km^{-1}, $f = 5.2$ GHz

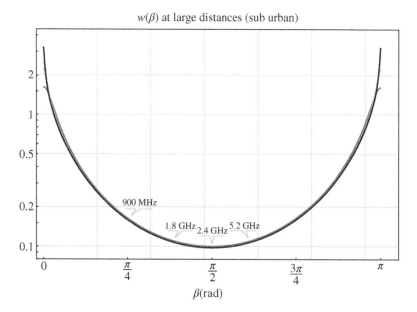

$w(\beta)$ at large distances (sub urban)

Figure 4.10 The PDF $\omega(\beta)$ in suburban areas far from the transmitter

As seen from Figure 4.10, with the decrease in the radiated frequency, the PDF, $\omega(\beta)$, smoothens in the range of inclination angles around 90°, and the ratio Δ becomes smaller. This tendency occurs both for suburban and urban areas with dense layout of the buildings and complicated buildings' profile with small "profile" parameters n.

4.5 Analysis of Polarization Ellipse Energetic Parameters

As was mentioned in the previous paragraph, we divided the areas under study into four different types: rural, mixed residential, suburban, and urban. This division yields large differences between the various built-up area characteristics. In order to evaluate every area parameter, we used data presented in [3, 4] for various rural, mixed residential, and urban areas and finally took the average of each parameter, as presented in Table 4.1, estimating the ranges of values of their variations.

4.5.1 The Ratio Δ vs. the BS Height

Figures 4.11 and 4.12 represent the ratio between the vertical and the horizontal components, as a function of the BS height, for distances of 500 and 1000 m between the Rx and Tx antennas operating with frequencies of

(a)

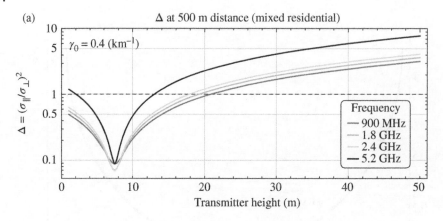

(b)

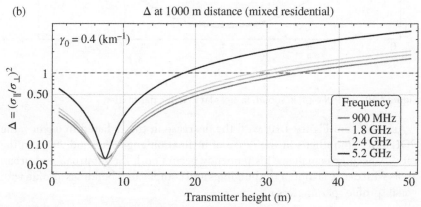

Figure 4.11 The Δ ratio vs. BS height (in m) in mixed-residential areas with $\gamma_0 = 0.4$ km^{-1} and operating frequencies of 900 MHz to 5.2 GHz: (a) distance of 500 m and (b) distance of 1000 m between the antennas

900 MHz, 1.8 GHz, 2.4 GHz, and 5.2 GHz in mixed-residential (Figure 4.11) and suburban (Figure 4.12) areas [3, 4]. The location of the horizontal axis represents the ratio $\Delta = \sigma_{\parallel}^2/\sigma_{\perp}^2 = 1$, which means that when the ratio is above the horizontal axis, the vertical component is smaller than the horizontal component. The situation when the ratio is below the horizontal axis indicates cases when the vertical component is bigger than the horizontal component.

From Figures 4.11 and 4.12, for every type of terrain, the minimal ratio is achieved when the BS antenna is located at the level of the average height of the buildings and when the ratio becomes smaller than the unit. In fact, when the BS antenna is located at the level of the average buildings' height, the horizontal and vertical components of the elliptical polarized field are roughly the same. With the changes of the BS antenna height to be lower (or higher) than the average height of the building's profile, the vertical component is decreased (or

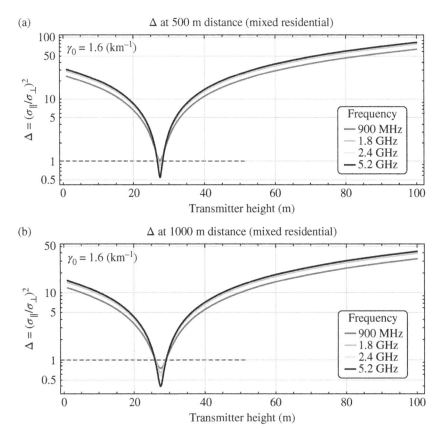

Figure 4.12 The Δ ratio vs. BS height (in m) in suburban areas with $\gamma_0 = 0.4$ km^{-1} and operating frequencies of 900 MHz to 5.2 GHz: (a) distance of 500 m and (b) distance of 1000 m between the antennas

increased), due to a decrease in signal losses caused by the multipath (stochastic) interference, and therefore, the ratio is also increased.

Next, for every type of terrain with an increase in the Tx–Rx distance, the ratio Δ decreases due to stronger signal losses in the vertical plane with respect to those in the horizontal plane, and as a result, with an increase in the Rx–Tx distance, the ratio also decreases. Moreover, from the presented illustrations, it is clear that for mixed-residential (and also rural) areas, the ratio Δ is smaller than 1, because in this terrain the BS height is low enough and due to stronger sporadic interference caused by multipath effects in the horizontal plane with respect to that in the vertical plane. As for the urban areas, it is seen that the ratio becomes larger than the unit because the transmitter height is much higher than the average height of the surrounding buildings. Here, conversely, the horizontal component is smaller than the vertical component, and the

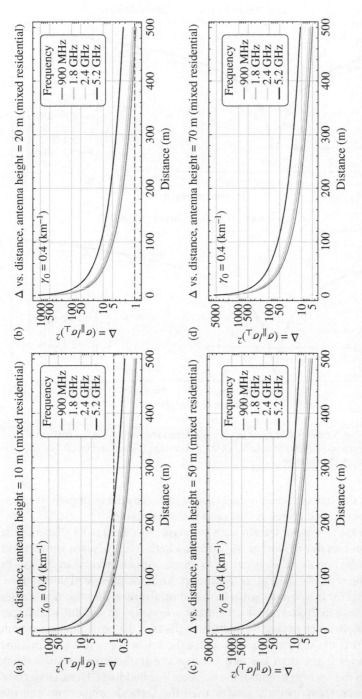

Figure 4.13 The Δ ratio vs. the distance between BS and MS in mixed-residential areas. BS antenna heights: (a) 10 m, (b) 20 m, (c) 50 m, and (d) 70 m with operating frequencies of 900 MHz to 5.2 GHz

building density becomes a more significant parameter of the multipath in the vertical plane due to multiple diffraction from the building roofs in the vertical plane, with respect to multiple scattering from the building walls in the horizontal plane.

4.5.2 The Δ Ratio vs. the Distance Between BS and MS Antennas

Figure 4.13a–d represents the ratio between the vertical and the horizontal components as a function of the distance between the BS and MS antennas for mixed-residential areas up to a distance of 500 m. Four cases are calculated for the heights of the BS equal 10 m, 20 m, 50 m, and 70 m for frequencies of 900 MHz, 1.8 GHz, 2.4 GHz, and 5.7 GHz, according to [3, 4].

The same calculation is repeated for suburban areas, where, in general, buildings are higher than in mixed-residential areas (both the lowest and the highest buildings). Therefore, the BS antennas are positioned higher than in the mixed-residential case with heights of 20 m, 50 m, 70 m, and 100 m. The results are shown in Figure 4.14a–d.

For rural areas, Δ < 1, in the cases of low BS antenna and relatively far from the BS. This means that the horizontal component is bigger than the vertical component compared with built-up areas, where most of the curves exceed unit. With the decrease in transmitter height (limiting to the receiver height), Δ < 1. This is evident, since the vertical component becomes smaller than the horizontal component, due to much stronger interference in the vertical plane. Moreover, when the transmitter achieves the average height of the building profile, lower line-of-sight condition is observed in the vertical plane causing stronger interference and the corresponding radio signal loss. When the distance between the receiver and the transmitter is short, the observed interference becomes weaker, resulting in an increase in the vertical component of the elliptically polarized wave, and the ratio Δ increases. As the distance between the transmitter and the receiver is increased, the ratio Δ decreases due to the increase in the sporadic interference phenomena caused by the strong multipath effect in the vertical plane (called the randomization of the vertical component of the elliptically polarized wave).

4.6 Analysis of the Loss Characteristics

We analyze each component of the elliptically polarized radio wave and present below the loss (in dB) for both components separately vs. the buildings' density and additionally vs. the distance between Rx and Tx. Computations were carried out for each type of the terrain with its specific parameters shown in Table 4.1 and for the operating carrier frequency of the BS of 2.4 GHz.

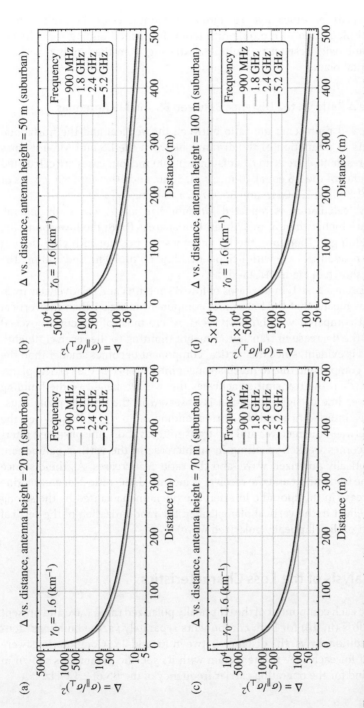

Figure 4.14 The Δ ratio vs. the distance between BS and MS in suburban areas. BS antenna heights: (a) 20 m, (b) 50 m, (c) 70 m, and (d) 100 m with operating frequencies of 900 MHz to 5.2 GHz

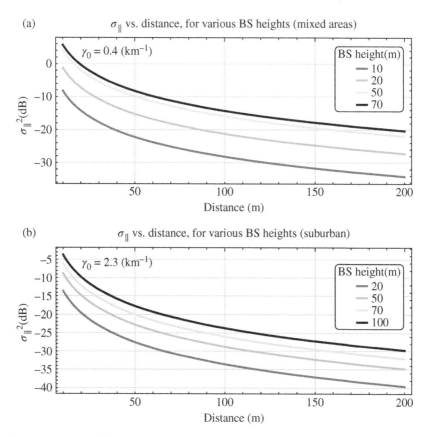

Figure 4.15 Loss of the horizontal component vs. the distance between the antennas for various BS heights in mixed-residential areas (a) and suburban (b) areas

4.6.1 Horizontal Component of the Total Elliptically Polarized Field

The results of computations are shown in Figure 4.15. One can see that in mixed-residential areas, the loss is smaller than in the suburban areas (the difference is roughly ~ 5 dB) mainly due to signal power loss. In suburban areas, the depolarization loss is higher, mainly due to higher density of the buildings' surrounding both the terminal antennas, Rx and Tx. As expected, the loss decreases with the increase in BS height – lower antennas demonstrate higher loss of ~ 10 dB.

4.6.2 Vertical Component of the Total Field

Irrelevant of the model used, $\sigma_\perp$ is independent of the heights of the BS and MS, as follows from expressions (4.17), (4.29), and (4.31). The corresponding

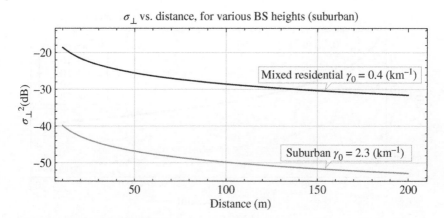

Figure 4.16 Loss of the vertical component vs. the distance between the MS and BS in the mixed-residential and suburban areas

results of computations are shown in Figure 4.16. For each terrain type one representative building density was selected, and again, the operational frequency is 2.4 GHz.

Figure 4.16 shows that when the distance between the transmitter (BS) and the receiver (MS) increases, the loss in the vertical component is apparent. In suburban and mixed-residential areas, the distance between the transmitter and the receiver may be increased, in order to cover more ground, because the loss is relatively reasonable. We note that in purely urban areas, one cannot increase the distance between the transmitter and the receiver since the loss is much more significant: in such areas, we need to bring the transmitter and the receiver close to each other, in order to decrease signal loss in accordance with the area of the users' service.

4.7   Path Loss Factor Due to Depolarization Phenomena

The angle of depolarization Υ can be defined as (usually it is taken to be positive [3, 4])

$$\Upsilon = \left| \sin^{-1} \frac{\sqrt{\sigma_{\|}^2}}{\sqrt{\sigma_{\text{total}}^2}} \right| = \left| \sin^{-1} \sqrt{\frac{\sigma_{\|}^2}{\sigma_{\perp}^2 + \sigma_{\|}^2}} \right| \tag{4.37}$$

The parameter Υ, usually called *antenna cross-polarization discrimination*, plays an important role in determining the propagation channel or the corresponding antenna performance [3]. Υ determines the corresponding polarization loss factor (PLF), which is used as a figure of merit to measure the degree of

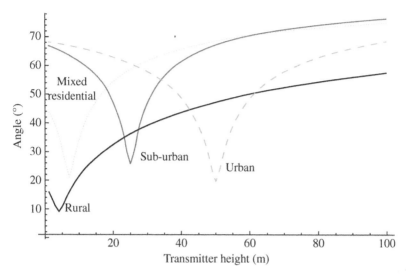

Figure 4.17 Depolarization angle vs BS height for rural, mixed-residential, suburban, and urban environments for distance of 500 m between antennas

polarization mismatch. The PLF is defined through the angle of depolarization as [3]

$$\text{PLF} = |\cos \Upsilon|^2 \tag{4.38}$$

To investigate the angle of depolarization, and the corresponding PLF, we examined, for mixed-residential and suburban environments, several scenarios. In each scenario, the distance between the antennas was fixed, and we varied the height of the BS antenna. Distance was taken $L = 500$ m between antennas. Figure 4.17 shows the behavior of the depolarization angle. For both cases, the depolarization angle reaches its minimal value when the BS height is equal to the average buildings' heights $\overline{h}$ where a sharp decrease in the depolarization angle is noticed. Further analysis showed that the larger the distance between antennas is, the lower the depolarization angle gets at BS.

The behavior of the PLF follows immediately from its definition in expression (4.38). Computations according Eq. (4.38) are presented in Figure 4.18a,b for mixed-residential and suburban areas, respectively, parameters of which were taken from Table 4.1. As follows from Figure 4.18a,b, when the BS antenna is positioned above the average height of buildings (i.e. $z_2 > \overline{h}$, when the sharp decrease in Υ is noticed), the angle of depolarization and PLF increase significantly. Notice a similar behavior with the increase in the ratio Δ. It is evident that in urban and suburban environments, due to multiple diffractions from building roofs and corners, the vertical component of the total field intensity variation, $\sigma_{\parallel}$, exceeds the horizontal component $\sigma_{\perp}$. At the same time, the

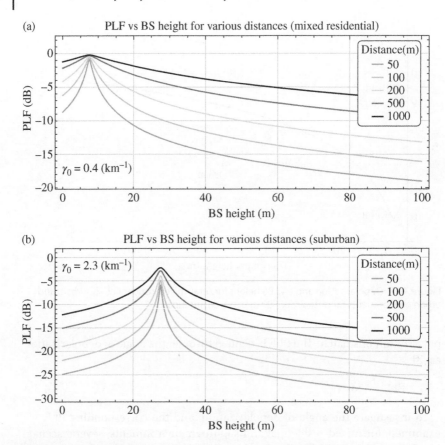

Figure 4.18 Power-loss factor (PLF) vs. BS height for mixed-residential (a) and suburban (b) environments for various distances between the antennas

PLF can achieve small values in the urban and suburban areas, since deterministic multidiffraction processes from tall buildings occurring in the vertical plane give the predominant impact in total depolarization loss, whereas the effects of random multiscattering process occurring in the horizontal and vertical planes can be mitigated. As for mixed-residential (and also rural) areas, the angle of polarization mismatch is smaller, the PLF can achieve higher magnitudes, and the horizontal component becomes to be prevailed, indicating the importance of the multipath phenomena due to multiple scattering from obstructions located in the horizontal plane [4]. Simultaneously, a decrease in angle of depolarization (see Figure 4.17) was accompanied by the increase of the ratio Δ and decrease in PLF (being negative in dB, see Figure 4.18a and b).

This peculiarity can also be explained by the increasing of the role of the horizontal component with respect to the vertical component of the polarization ellipse. This means that in the horizontal plane, random processes of multipath become predominant, and the PLF parameter can achieve in urban and suburban environments magnitudes of around -10 dB, whereas in mixed-residential and rural areas, its magnitude increases drastically achieving even -20 dB, indicating the role of random multipath processes, such as multiple scattering from obstructions, in total wave field depolarization. Simultaneously, a decrease in the angle of depolarization was accompanied by the increase in the ratio Δ and decrease in the PLF (being negative in dB).

4.8 Conclusions

In order to predict the influence of depolarization on propagation of the elliptically polarized radio wave, it is necessary to obtain information on the main characteristics and parameters of the terrain. In fact, each of the discussed terrain types acts like a communication channel that "reacts" differently on the input propagation parameters, that is, on the propagation environment within each channel: urban, suburban, mixed residential, and rural.

The channel "reaction" depends on different terrain factors: antenna location and elevation with respect to the buildings' heights, obstruction characteristics (e.g. the permittivity of the material – stone, wood, steel, glass, etc.), buildings' density, distance between the transmitter and the receiver, degree of roughness of the walls, buildings' width or length, terrain topography, and so on.

The formulas that describe the intensity distribution of the elliptically polarized radio wave inside the ellipse, which until recently were not presented in the literature, were derived based on the main formulas of signal intensity distribution in space domain based on the multiparametric stochastic theoretical framework that describes radio propagation in various terrain environments.

This chapter represents a novel approach of developing expressions for wave depolarization of the incoherent part I_{inc} in mixed-residential and suburban areas, in terms of the parameters of the stochastic model presented in [3, 4]. The simple "engineering" formulas for radio wave intensity deviations in the vertical and horizontal planes of the ellipse, the corresponding angle of depolarization as function of its vertical and horizontal components, and the PLF were derived based on the proposed stochastic approach.

Expressions for the ratio $\Delta = (\sigma_{||}/\sigma_{\perp})^2$ and for the distribution of the inclination angle β are given for each terrestrial area type. This allows one to predict the behavior of the inclination angle of the polarization ellipse, as well as the distribution of the energy in the parallel and perpendicular planes of the polarization ellipse. Moreover, the additional investigation of the full set of PDF functions, $\omega(\theta)$, $\omega(\beta)$, and $\omega(R)$, given in expressions (4.34)–(4.36) for a much larger set of parameters becomes to be important in dense built-up areas, since the power received from the non-LOS component (i.e. I_{inc}) is a much more predominant part of the received wave.

The corresponding 3D numerical code was performed for analysis of the corresponding formulas for various terrain scenarios, urban, suburban, mixed-residential, and rural, depending on the built-up terrain features. Depolarization effects and polarization losses were analyzed for two types of environments: mixed-residential and suburban.

The obtained results allow us to state additionally that:

1. In mixed-residential areas, the vertical component of the elliptically polarized radio wave is not changed significantly (e.g. has small depolarization loss), and the angle of depolarization is too small with respect to that obtained in the urban and suburban areas. This allows us to suggest the increase in the range between the transmitting (Tx) and the receiving (Rx) antenna in such areas.

2. In suburban areas, the wave intensity loss is significant both in the vertical and the horizontal planes of the elliptically polarized wave, caused by the random interference of its multipath components due to multiscattering, multidiffraction, and multireflection phenomena from obstructions surrounding both the terminal antennas.

3. As expected, the angle of depolarization is larger for suburban (and urban) channels with the corresponding increase in the PLF. This effect strongly depends on the height of the transmitter antenna (the receiver antenna was always lower than the building's roofs) with respect to the average building's height. Thus, the angle of depolarization decreases with a decrease in the transmitter antenna height and vice versa – with an increase in the transmitter antenna height.

4. The ratio between the vertical and the horizontal components of the elliptically polarized wave increases with the transfer of the channel from rural to urban scenarios. This means that the effect of depolarization becomes more significant in the more dense built-up environments.

5. Increase in the depolarization of the even elliptically polarized radio wave, passing through the suburban and urban channels, yields the increase in the randomization of the wave intensity both in the vertical and horizontal planes, leading to changes in the shape of the ellipse and its rotation in a large angle.

6. Additionally, the increase in the depolarization loss and the angle of depolarization yield the decrease in the signal power and require additional signal amplification at the receiver.

7. Knowledge of the "reaction" of each individual channel (urban, suburban, mixed residential, and rural) on signal depolarization allows to give for each designer of wireless communication links a powerful tool for predicting a priori the influence of the built-up channel "response" on the depolarization phenomena accounting for each specific scenario occurring at the built-up scene.

References

1 Ponomarev, G.A., Kulikov, A.N., and Telpukhovsky, E.D. (1991). *Propagation of Ultra-Short Waves in Urban Environments*. Tomsk, Rasko: USSR.

2 Yakovlev, O.I., Yakubov, V.P., Uryadov, V.P., and Pavel'ev, A.G. (2009). *Propagation of Radio Waves*. Moscow: Russian Publisher.

3 Blaunstein, N. and Christodoulou, C. (2014). *Radio Propagation and Adaptive Antennas for Wireless Communication Networks – Terrestrial, Atmospheric and Ionospheric*, 2e. New Jersey: Wiley.

4 Ben-Shimol, Y., Blaunstein, N., and Sergeev, M.B. (2015). Depolarization effects in various built-up environments. *Sci. J. Inf. Control Syst.* 69 (2): 83–94.

5

Theoretical Framework for Positioning of Any Subscriber in Land–Land and Atmosphere–Land Multiuser Links

As was mentioned in the previous chapter, to mitigate the effects of multiplicative noise and the noise due to interference between users in multiple access communication, the directional, sectorial, and adaptive antennas (phased array or multibeam) are used in one or both ends of the wireless channel, which was defined in Chapter 2 as a multiple-input-multiple-output (MIMO) channel. Using the multibeam adaptive antenna together with the corresponding processing algorithms operating in the space, time, and frequency domains (see details in Refs. [1, 2] and references therein) allows the channel to radiate the desired energy in the desired direction or to cancel the undesirable energy from the undesirable direction (see Chapter 3). The same method is used to minimize the effect of multipath fading. However, the operational ability of adaptive antennas strongly depends on the degree of accuracy to localize any subscriber located in the areas of service with high layout of buildings and huge amount of users simultaneously integrated into multiuser assess [3–6]. And this problem during the recent two to three decades was the main, among others, in deployment of future generation of wireless networks beyond 3G. For these purposes, the corresponding propagation models can predict not only distribution of each signal energy in space domain but also the angular, azimuth and elevation, time-delay, and frequency distribution of the multipath components of the total signal arriving at each user's receiver [1, 2, 7–15]. As was mentioned in Chapter 3, following [5, 6], several realistic channel models were created at the beginning of the twentieth century for adaptation of various-type multibeam antenna in real operation environment, land and atmospheric, and for the estimation of the obtainable capacity gain [1, 7–10]. In other words, to design effectively or improve existing wireless networks with optimal user and frequency allocation and cellular and noncellular planning, the wireless network designers need realistic spatial and temporal channel models combined with high-resolution precise experiments. The latter are required for "parameterization and validation" of such channel models using various probabilistic theoretical and experimentally obtained frameworks [1, 2, 11–15]. Despite the importance of these aspects, a limited number of high-resolution spatial and temporal experiments are available, and only few of them use

Advanced Technologies and Wireless Networks Beyond 4G, First Edition.
Nathan Blaunstein and Yehuda Ben-Shimol.
© 2021 John Wiley & Sons, Inc. Published 2021 by John Wiley & Sons, Inc.

three-dimensional (3D) measurements accounting for not only angle-of-arrival (AOA) and time-of-arrival (TOA) but also elevation-of-arrival (EOA) and frequency (Doppler shift, DS) distributions of the multipath components within the communication channel (see Refs. [1, 2] and references there). The same situation also occurs with theoretical prediction of these characteristics; only few models, simple or more complicated, exist, which can describe the mutual AOA and TOA distributions of the total signal, and at the same time may account different situations in the corresponding environment [1, 2, 5–15], namely, in the urban scene a predicting model must account for various built-up parameters such as height, density, and real street orientation of a building as well as position of the antenna with respect to the overlay profile of the buildings (see Chapter 3). Furthermore, as was mentioned in Chapter 3, the network's gain is another important issue that has high priority in cellular network performance.

The analysis of parameters, which influence the correlation of data stream parameters, the capacity and the spectral efficiency, of each user's channel, is based on an understanding of signal AOA and TOA distribution, which are based on the corresponding theoretical models (see Refs. [1, 2, 5–15]). Figure 5.1 shows the main concept of spatial, AOA (azimuth, φ), and TOA (time delay, τ) signal distribution to obtain exact position of each subscriber during multiuser communication access (taken according to [6]).

In Section 5.1, we briefly present the stochastic multiparametric model that was proved by numerous experimental and theoretical investigations of the propagation problem in built-up areas: rural, mixed-residential, suburban, and urban, with different building and other obstruction profile and density layouts (see Refs. [1, 2, 5–15] and references herein). Here, we present the reader some useful recommendations on how to predict a signal power in

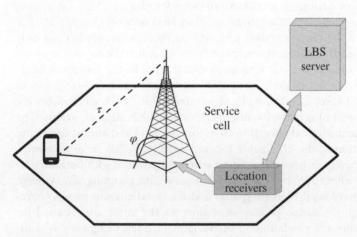

Figure 5.1 Schematic presentation of any subscriber localization by knowledge of the azimuth (φ) and time-delay (τ) determination (according to [2])

spatial (azimuth and elevation), temporal (TOA), and frequency (DS) domains based on modified stochastic approach proposed in Refs. [1, 2, 5–15].

These aspects are very actual for localization of any desired subscriber in the area of service, both land-to-land and land-to-atmosphere. Then, in Section 5.2, we show how localization of any subscriber in the land and atmospheric environments can be predicted using this model for specific dense buildings' overlay profile and their dense layout and based on the proposed probabilistic models following Refs. [1, 2, 7–15].

5.1 Signal Power Distribution in the Space, AOA, TOA, and Frequency Domains for Prediction of Operative Parameters of Sectorial and Multibeam Antennas

5.1.1 Signal Intensity Distribution in Space Domain. According to 3-D Stochastic Approach

Following [2, 3], we can obtain the total average signal intensity in the space domain, that is, in $\mathbf{r}$-domain. The average intensity of the received field $\langle I(\mathbf{r}_2) \rangle = \langle U(\mathbf{r}_2) \cdot U^*(\mathbf{r}_2) \rangle$ can be presented as

$$\langle I(\mathbf{r}_2) \rangle = 4k^2 \int_{S_B} dS_B \int_{S_B} dS_B' \cdot K(\mathbf{r}_{S_B}, \mathbf{r}_{S_B}') \cdot G(\mathbf{r}_2, \mathbf{r}_{S_B}') \cdot$$
$$G(\mathbf{r}_2, \mathbf{r}_{S_B}) \cos \psi_{S_B} \cos \psi_{S_B}' \tag{5.1}$$

where $K(\mathbf{r}_{S_B}, \mathbf{r}_{S_B}')$ is the correlation function of the total field at points $\mathbf{r}_{S_B}$ and $\mathbf{r}_{S_B}'$ located at surface S_B of arbitrary building, whereas the source is located at the point $\mathbf{r}_1$ [6, 7]:

$$K(\mathbf{r}_{S_B}, \mathbf{r}_2') = 4k^2 \left\langle \int_{S_B} dS_B \int_{S_B} dS_B' \cdot Z(\mathbf{r}_2, \mathbf{r}_{S_B}, \mathbf{r}_1) \cdot Z(\mathbf{r}_2', \mathbf{r}_{S_B}', \mathbf{r}_1) \cdot \right.$$
$$\Gamma^*(\varphi_{S_B}', \mathbf{r}_{S_B}') \cdot \sin \psi_{S_B} \sin \psi_{S_B}' \cdot G(\mathbf{r}_2, \mathbf{r}_{S_B}) \cdot G(\mathbf{r}_{S_B}, \mathbf{r}_1) \cdot$$
$$\left. G^*(\mathbf{r}_2', \mathbf{r}_{S_B}') \cdot G^*(\mathbf{r}_{S_B}', \mathbf{r}_1) \right\rangle \tag{5.2}$$

Here, the reflection coefficient is given by

$$\Gamma(\varphi_S, \mathbf{r}_S) = \Gamma \exp\left(-\frac{\zeta}{\ell_v}\right)$$

and the shadow function $Z(\mathbf{r}_2, \mathbf{r}_1)$ is a superposition of all probability functions of line of site (LOS) between any subscriber and the base station (BS). By averaging (5.1) over the spatial distribution of the nontransparent screens, their number, and the reflection properties of these screens, we obtain the following formula for the single-scattered field (single diffraction from the building's rooftops) [6, 7] (see also Chapter 3):

$$\langle I_{\text{inc}}(\mathbf{r}_2) \rangle = \frac{\Gamma \lambda \ell_v}{8\pi[\lambda^2 + (2\pi\ell_v\gamma_0 F(z_1, z_2))^2]} \frac{\sqrt{\frac{\lambda d}{4\pi^3} + (z_2 - \overline{h})^2}}{d^3} \tag{5.3}$$

The same result can be obtained for the double-diffracted and double-scattered field as [6, 7]

$$\langle I_{\text{inc}}(\mathbf{r}_2) \rangle = \frac{\Gamma^2 \lambda^3 \ell_v^2}{24\pi^2[\lambda^2 + (2\pi\ell_v\gamma_0 F(z_1, z_2))^2]^2} \frac{\frac{\lambda d}{4\pi^3} + (z_2 - \overline{h})^2}{d^3} \tag{5.4}$$

The coherent part of the total field intensity can be obtained as [6, 7]

$$\langle I_{\text{co}}(\mathbf{r}_2) \rangle = \exp\left[-\gamma_0 d \frac{F(z_1, z_2)}{z_2 - z_1}\right] \frac{\sin^2(kz_1z_2)/d}{4\pi^2 d^2} \tag{5.5}$$

where the function of the buildings' overlay profile between two antennas was presented in Chapter 3.

Finally, the total average field intensity is written as

$$\langle I_{\text{total}} \rangle = \langle I_{\text{co}} \rangle + \langle I_{\text{inc}} \rangle \tag{5.6}$$

Hence, the total path loss (in dB) is presented as

$$L_{\text{total}} = 10\log_{10}[\lambda^3(\langle I_{\text{co}} \rangle + \langle I_{\text{inc}} \rangle)] \tag{5.7}$$

5.1.2 Signal Energy Distribution in Angle-of-Arrival (AOA) and Time-of-Arrival (AOA) Domains

It was proved during the recent decades, both experimentally and theoretically (see [4, 5] and references therein), that the azimuth and delay spread distributions of signal energy are independent, that is,

$$f(\varphi, \tau) = f_1(\varphi) \cdot f_2(\tau) \tag{5.8}$$

This fact was used in Ref. [5, 6] to obtain a joint PDF for modified stochastic multiparametric model. Thus, a stochastic approach was presented by use of the Fourier transform of the total correlation function described by Eq. (5.2), converting the presentation of the latter function in the Cartesian coordinates to the polar coordinates – the azimuth and the radius vector of any point along the radio path. Following this approach, we can present the power spectrum of a signal in the azimuth domain at the receiving antenna located at a point (see Figure 5.2) in the following form [7]:

$$W(\varphi) = \frac{\Gamma \lambda \ell_v \overline{h}}{16\pi^2[\lambda^2 + (2\pi\ell_v\gamma_0\overline{h})^2]d^3}[f_1(\varphi) + f_2(\varphi)] \tag{5.9}$$

where

$$f_1(\varphi) = \frac{2z_1^2(\gamma_0 d)^2}{z_2 + \overline{h}} \frac{\zeta'(1 - \cos\varphi)}{\overline{h}}$$

$$\frac{\exp\left[-\gamma_0 d\left(\dfrac{\overline{h}}{z_2} + \dfrac{\zeta'}{2} \dfrac{1+\cos\varphi}{1 + \frac{\gamma_0 d}{2}\left(1+\frac{\overline{h}}{2}\right)(1-\cos\varphi)}\right)\right]}{1 + \frac{\gamma_0 d}{2}\left(1 + \dfrac{\overline{h}}{z_2}\right)(1 - \cos\varphi)} \qquad (5.10)$$

$$f_2(\varphi) = \frac{2\overline{h}\gamma_0 d}{z_2 + \overline{h}}\left[1 + \frac{\overline{h}}{z_2}\frac{1 + (k\ell_v\gamma_0 d)^2}{1 + (\gamma_0\zeta'd)^2}\right]$$

$$\frac{\exp\left[-\gamma_0 d\left(\dfrac{\overline{h}}{z_2} + \dfrac{\zeta'}{2} \dfrac{1+\cos\varphi}{1 + \frac{\gamma_0 d}{2}\left(1+\frac{\overline{h}}{2}\right)(1-\cos\varphi)}\right)\right]}{1 + \frac{\gamma_0 d}{2}\left(1 + \dfrac{\overline{h}}{z_2}\right)(1 - \cos\varphi)} \qquad (5.11)$$

Here, Γ is the reflection coefficient from the building surface, ℓ_v is the height or width of the building segments (window, balcony, etc.), λ is a wavelength, z_1 is the height of the BS (point A), and z_2 is the height of the moving subscriber (point B), as shown in Figure 5.2.

Here, as in [6, 7], a new parameter is introduced

$$\zeta' = \frac{\sqrt{\dfrac{\lambda d}{4\pi^3} + (z_2 - \overline{h})^2}}{z_2}$$

Figure 5.2 The 2D geometry of single scattering from the nontransparent screen (building)

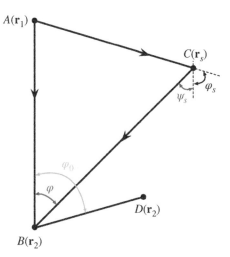

which accounts for the process of diffraction from buildings, instead of the $\zeta = (z_2 - \overline{h})/z_2$, used for 2-D model in [5], which did not take into consideration the diffraction phenomenon.

Expression (5.9) consists of two main terms f_1 and f_2, each relating to a different propagation phenomenon. The term f_2 is the significant one that describes the influence of the scattering area located at the proximity of the mobile subscriber (MS). f_1 describes the general effect of rare scatterers that are distributed uniformly in areas surrounding the BS and MS. The influence of different scatterers for the three typical cases, depending on the BS antenna height, is sketched in Figure 5.3, according to [7].

When both antennas are lower than the height of the buildings (see Figure 5.3c), then both components, f_1 and f_2, should be taken into account.

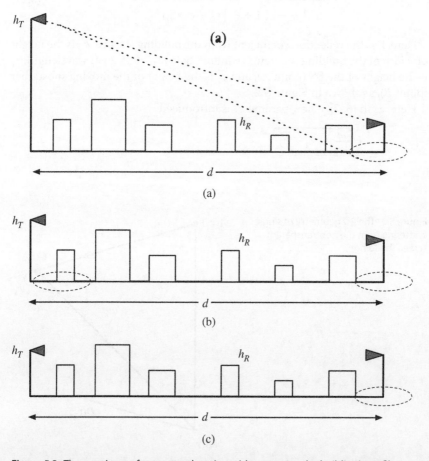

Figure 5.3 Three variants of antennas elevation with respect to the building's profile

From formulas (5.10) and (5.11), it follows that, if the base station antenna height increases up to $z_2 = \overline{h}$ (see Figure 5.3b), then $\zeta' \ll 1$ and $f_1(\varphi)|_{z_2=\overline{h}} > f_2(\varphi)|_{z_2=\overline{h}}$. In this case of $z_2 = \overline{h}$, $f_1(\varphi)$ is close to zero, and it means that all scatterers located in the far zone from the MS, near the BS, do not influence on spreading of the total signal at the BS.

With an increase in the height of the BS antenna, that is $z_2 > \overline{h}$, the influence of the buildings surrounding the MS on the total signal distribution will be more significant, and f_2 becomes larger than f_1, describing the effect of scatterers located close to the MS (see Figure 5.3a).

In our further computations presented below, we use the normalized function $\widetilde{W}(\varphi)$, which describes the average angle spectrum density of single-scattered waves with regard to the total average field intensity

$$\widetilde{W} = \frac{W(\varphi)}{\langle I \rangle} \tag{5.12}$$

where $\langle I \rangle$ is defined above by expression (5.6).

Finally, formula (5.12) can be deduced to the normalized signal energy form [9]:

$$\widetilde{W}(\varphi) = \frac{\overline{h}}{z_2}\left[1 + \left(\frac{\overline{h}}{z_2}\right)^2 \frac{1 + (k\ell_v\gamma_0\overline{h})}{1 + \gamma_0^2(\overline{h} - z_1)^2 \frac{\lambda d/4\pi^3 + (z_2-\overline{h})^2}{z_2^2}}\right]f_1(\varphi) +$$

$$\frac{\sqrt{\lambda d/4\pi^3 + (z_2 - \overline{h})^2}}{\overline{h} - z_1}f_2(\varphi) \tag{5.13}$$

The first term consists of the function $f_1(\varphi)$, which can now be presented as

$$f_1(\varphi) = \frac{\gamma_0 d}{2}\frac{\exp[-\gamma_0 d(1 - \widetilde{\zeta})]}{1 - \widetilde{\zeta}/2}\frac{\exp\left[-\frac{\gamma_0 d}{2}\frac{\widetilde{\zeta(1+\cos\varphi)}}{1 + \gamma_0 d(1-\widetilde{\zeta}/2)(1-\cos\varphi)}\right]}{1 + \gamma_0 d(1 - \widetilde{\zeta}/2)(1 - \cos\varphi)} \tag{5.14}$$

and the second part consists of $f_2(\varphi)$, which for the case of $\pi/2 < |\varphi| < \pi$ can be expressed as

$$f_2(\varphi) = \frac{1 - \cos\varphi}{2\gamma_0 d} \tag{5.15}$$

which does not depend on situations with elevations of both antennas, base station (BS) and mobile subscriber (MS), with respect to the building roof tops. In formulas (5.13)–(5.15), the average height $\overline{h}$ is used for computations instead of $F(z_1, z_2)$, which corresponds to 2-D model, where, as was shown in [6, 7], the building's profile is not so actual in AOA and TOA domains. In the 2-D case, the parameter $\widetilde{\zeta}$ can be obtained in a more simplified form $\widetilde{\zeta} = (\overline{h} - z_1)/(z_2 - z_1)$, [1, 4]. The signal energy distribution in time-delay domain can be obtained in the same manner [2, 3]:

$$W(\tau) = \frac{\Gamma}{8\pi^2 d^2} \frac{k\ell_v \gamma_0 \overline{h}}{1 + (k\ell_v \gamma_0 \overline{h})^2} \left\{ 1 + (1 - \zeta') \left[1 + (1 - \zeta')^2 \frac{1 + (k\ell_v \gamma_0 \overline{h})^2}{1 + (\zeta' \gamma_0 \overline{h})^2} \right] \right.$$

$$\left. f_1(\tau) + \frac{\zeta'}{1 - \zeta'} f_2(\tau) \right\} \tag{5.16}$$

where

$$f_1(\tau) = \frac{(\gamma_0 d)^2 \sqrt{\tau^2 - 1}}{4\tau^2} \exp\left[-\gamma_0 \tau \frac{2 - \zeta'}{2} d \right] I_0 \left(\frac{\gamma_0 \zeta' d}{2} \right) \tag{5.17}$$

$$f_2(\tau) = \frac{\gamma_0 d}{2} \exp\left(-\frac{\gamma_0 \tau d}{2} \right) \left[\exp\left(-\frac{\gamma_0 \tau d}{2} \right) + \frac{\sqrt{\tau - 1}}{\sqrt{\tau + 1}} I_0 \left(\frac{\gamma_0 d}{2} \right) \right] \tag{5.18}$$

In the time-delay domain, the same properties of the signal power distribution, as discussed above for the azimuth domain by use of Figure 5.3, can be easily found. The function from (5.16) satisfies the following condition:

$$\int_0^\infty W(\tau) d\tau = \langle I \rangle \tag{5.19}$$

Finally, we can normalize this spectrum and analyze $\hat{W}(\tau)$ of the time delay (with respect to the total field intensity taken from (5.6)). As was analyzed, with the increase of the antenna height z_2 over the building's layer, the probability of the direct visibility between the receiver and the neighborhood of the transmitter, which has the maximal luminance, increases. As a consequence, the energetic spectrum of the time delay can be presented and derived for $\zeta > 0.2$ (i.e. $z_2 - \overline{h} > 2$ m) by the following formula:

$$W(\tau) = \frac{\langle I \rangle f_2(\tau)}{\int_0^1 f_2(\tau) d\tau} \tag{5.20}$$

The integral scale of the normalized signal power $W(\tau)/\langle I \rangle$ for the case when the maximum of $f_2(\tau)$ at the point $\tau = 1 + 1/(\gamma_0 d)$ approximately equals $\gamma_0 d/3$ and for the case $\zeta > 0.2$ and $\gamma_0 d > 6$ can be easily estimated as

$$\Delta \tau_s = 3\gamma_0/d \tag{5.21}$$

Hence, the same features of signal power spectrum distribution, as we found above in the AOA domain, were observed in the time domain with change of base station antenna height, ranges between both terminal antennas and environmental conditions in the urban scene.

5.1.3 Signal Power Spectrum in the Frequency and Doppler-Shift (DS) Domains

Let us now present signal energy distribution in the frequency domain following [6, 12–14]. In built-up areas, the spatial distribution of signal strength fully determines the properties of temporal signal distribution obtained at the receiver, and the energy spectrum of the signal temporal deviations, which are created by the interference of waves arriving at the receiver, relates to the spectrum of spatial variations of signal strength through the following relationship [5, 6]:

$$\widetilde{W}(\tau, \varphi_0) = \frac{1}{v} \widetilde{W}\left(\frac{\omega}{v}, \varphi_0\right) \tag{5.22}$$

Here, v is the velocity of the transmitter (or receiver) traveling from point B to point D during the time $\tau = \ell/v$. The velocity v is related to the maximal Doppler frequency f_{dm} via $v = f_{dm}c/f$, where f is the radiated frequency; c is the speed of the light; and φ_0 is the angle between the direction of movement of the mobile antenna and the direction from point B to point A (see Figure 5.2). Using this relation between the signal spectra in the space and time domains, we can examine their distribution in various built-up areas with randomly distributed buildings. Let us consider two typical situations in the urban scene. Accounting for the case that the receiving antennas are above the built-up area ($z_2 \gg \overline{h}$, as shown in Figure 5.3a), we can use the 2-D approximation according to [6, 12–14]. In the 2-D case, the stationary transmitter antenna is at the *roofs level or below it* (as shown in Figure 5.3b,c, respectively). In the 2-D case one can obtain the following expression for the spectral function of signal temporal fluctuations:

$$\widetilde{W}(\omega, \varphi_0) = \frac{2 \sinh \chi \cdot \omega_d}{\sqrt{\omega_d^2 - \omega^2}} \frac{\omega_d \cdot \cosh \chi - \omega \cdot \cos \varphi_0}{(\omega_d \cdot \cosh \chi - \omega \cdot \cos \varphi_0)^2 - (\omega_d^2 - \omega^2)\sin^2 \varphi_0} \tag{5.23}$$

where $\omega_d = k \cdot v = \omega v/c = 2\pi f_{dm}$, $\sinh \chi$ and $\cosh \chi$ denote hyperbolic sine and cosine, respectively, and the parameter χ accounts for the density of buildings and the range between the antennas

$$\chi = \ln\left[\left(1 + \frac{1}{\gamma_0 d}\right) - \left(\left(1 + \frac{1}{\gamma_0 d}\right)^2 - 1\right)^{1/2}\right] \tag{5.24}$$

The frequency dependence of power spectrum, described by expression (5.23), is much more general with respect to that obtained by Clark in his 2-D statistical model [5, 6] and depends on several of the built-up terrain factors, such as the parameter γ_0, direction of moving vehicles (on φ_0), and their speed v (i.e. on $\omega_d \propto v/\lambda$). Really, according to Clark, the normalized Doppler power spectrum following can be presented in our notations as [5, 6]

$$\widetilde{W} = \frac{1}{\pi f_{dm}\sqrt{1 - \left(\frac{f}{f_{dm}}\right)^2}} \tag{5.25}$$

The same "band-limited" power spectral density (PSD) distribution has been obtained by Aulin in the so-called quasi 3-D model (sometimes also called in the literature "the 2.5-D model"), where the Doppler spectrum does not cover the full Doppler bandwidth $f_d \in [-f_{dm}, f_{dm}]$. The latter approximate model is usually used in the more realistic cases for mobile communication links whose antenna heights are less than the heights of the local obstructions surrounding. In this case the majority of the arriving multipath rays travel in the vertical plane in a nearly horizontal direction, with the elevation angles θ_i having a mean value close to $0°$.

To return to the 3-D case, we should notice that the coherent component of the total field intensity simply corresponds to an energetic spectrum with a single spectral line of $\varphi = 0$. We therefore can present the total energetic spectrum in the AOA domain for the total field intensity within the built-up layer as [9]

$$W(\varphi) = \langle I_{co} \rangle \delta(\varphi) + \langle I_{inc} \rangle \tag{5.26}$$

where $\delta(\varphi)$ is the Kronecker delta function, which equals 1, if $\varphi = 0$, and equals 0, if $\varphi \neq 0$. The dependences of both incoherent and coherent components of the total field as functions of antenna heights, distance between antennas and the building's spatial density, were described by (5.4) and (5.5), respectively. Here, we presented the incoherent component of the total field intensity as a sum of two separate components corresponding to "far" located obstructions and "in proximity" of the transmitting (subscriber) antenna, respectively, as was done above for $\widetilde{W}(\tau)$, that is,

$$\langle I_{inc} \rangle = \langle I_1 \rangle + \langle I_2 \rangle \tag{5.27}$$

$$\langle I_1 \rangle \approx (1 - \zeta) \left[1 + (1 - \zeta)^2 \frac{1 + (k\ell_v \gamma_0 \overline{h})^2}{1 + (\zeta \gamma_0 (\overline{h} - z_1))^2} \right] \cdot \frac{(\gamma_0 d)^2 \exp[-\gamma_0 d(1 - \zeta)]}{2 + \gamma_0 d} \frac{1 - \cos \Delta \widetilde{\varphi}_1}{\sin \Delta \widetilde{\varphi}_1} \tag{5.28}$$

and

$$\langle I_2 \rangle \approx \zeta \frac{1 - \cos \Delta \widetilde{\varphi}_2}{\sin \Delta \widetilde{\varphi}_2} \tag{5.29}$$

In the same manner can be presented the total energetic spectrum for the total field intensity in the time-delay domain:

$$W(\tau) = \langle I_{co} \rangle \delta(\tau) + \langle I_1 \rangle \widetilde{W}_1(\tau) + \langle I_2 \rangle \widetilde{W}_2(\tau) \tag{5.30}$$

The functions $\widetilde{W}_1(\tau)$ and $\widetilde{W}_2(\tau)$ in expression (5.30) are the normalized energy distribution of the signal in the TD domain, described correspondingly by two summands in (5.17) and (5.18), respectively.

Accounting for the relations between signal energy distribution in the frequency domain and in the Doppler-shift (DS) domain (5.22), we can present the same energy distribution for the bandpass radio signal in the same manner, as was presented by (5.26) and (5.30) in AOA and TOA domains.

Indeed, after direct Fourier transform and accounting for relation (5.30), after straightforward computations, we obtained the following signal normalized spectrum in the DS domain [6, 13]:

$$\widetilde{W}(f) = \frac{K}{K+1}\delta(\text{LOS}) +$$

$$\frac{1}{K+1}\frac{\exp\left(\kappa\frac{f}{f_{dm}}\cos(\varphi_0)\right)\cosh\left(\kappa\sqrt{1-\left(\frac{f}{f_{dm}}\right)^2}\cos(\varphi_0)\right)}{\sqrt{1-\left(\frac{f}{f_{dm}}\right)^2}I_0(\kappa)}$$

$$(5.31)$$

where K is the Ricean parameter of frequency-selective fading caused both by the multiray effect and by the Doppler effect of moving receiver (see Figure 5.2), which was already presented and analyzed numerically. In expression (5.31) in the function $\cosh(\cdot)$ the parameter κ defines the antenna directivity and equals 0, if the antenna is isotropic, and 1, if the antenna is fully anisotropic with narrow beams; other parameters were introduced above following Refs. [6, 13].

5.2 Localization of Any Subscriber in Land Built-Up Areas

5.2.1 3-D Stochastic Model for Different Scenarios of Buildings' Layout

We consider two scenarios, I and II, of buildings' layout. One of them was discussed in [1, 11] and presented in Figure 5.4, which shows a representative example of the simulated urban scenario according to [1, 11]. The size of

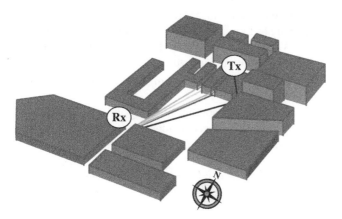

Figure 5.4 3D layout rearranged from [1, 11], accounting for buildings' profile

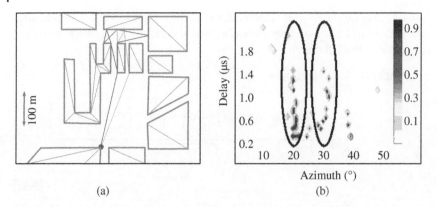

Figure 5.5 (a) The urban environment and (b) the collected measurements

the simulated propagation area was set to 900×900 m, and it consists of the randomly oriented buildings, arrays of buildings, and straight-street crossings of Manhattan's street plan. The building height was considered to be uniformly distributed with the mean height of 15 m.

The omnidirectional antenna with 4 dBi gain for the MS and a directional antenna with horizontal beamwidth of $120°$ and 17 dBi gain for the BS were simulated in [1, 11]. A dynamic range of 150 dB was considered for an appropriate balance between the transmitting power and the receiver sensitivity.

The second scenario was calibrated using the collected data from those obtained during the measurement campaign in Helsinki, Finland [1, 11], and is shown in Figure 5.5, where the BS antenna was placed below the rooftop level, $h_R = 10$ m. This figure shows a good match between the measured and collected data [1, 11], recorded due to guiding effects of two parallel streets near the low-height arranged antenna.

In Chapter 3, we introduced the building contours density, $\gamma_0 = 2\overline{L}\nu/\pi$ km^{-1}, where ν is the building density in 1 km^2. Considering, according to topographic maps, shown in Figures 5.4 and 5.5, the uniform distribution of the buildings' heights, the building height profile is characterized by the average building height, $\overline{h} = (h_1 + h_2)/2$, where h_1 and h_2 are the minimum and the maximum heights of the profile.

Then, the distribution of the nontransparent buildings in polar coordinates with a BS at its origin can be obtained when considering different heights of the BS antenna, h_R [6, 7]: For low-elevated BS and MS antennas:

$$\mu_1(\tau) = \pi \nu \gamma_0 d^3 (1 - \zeta) \tau \sqrt{\tau^2 - 1} \exp\left(-\gamma_0 \tau \frac{2 - \zeta}{2} d\right) I_0 \left(\frac{\gamma_0 \zeta d}{2}\right) \quad (5.32)$$

For high-elevated BS antenna and low-elevated MS antenna:

$$\mu_2(\tau) = 2\pi v \gamma_0 d^3 \frac{\zeta}{1-\zeta} \left[\tau \exp(-\gamma_0 \tau d) + \right.$$

$$\left. \frac{(\tau-1)^2}{\sqrt{\tau^2-1}} \exp\left(-\frac{\gamma_0 \tau d}{2}\right) I_0\left(\frac{\gamma_0 d}{2}\right) \right] \tag{5.33}$$

Here, $I_0(\cdot)$ is the zero-order modified Bessel function of the first kind; the parameter ζ was introduced earlier as $\zeta = (z_2 - \overline{h})/z_2$. The above expressions were obtained for the condition $\gamma_0 d \gg 1$ by taking into account the limit on the base station antenna height z_2 of $z_2/\overline{h} < \gamma_0 d$. Here, $\widetilde{r} = (d^2 + r^2 - 2rd \cos\phi)^{1/2}$, and r is defined in Figure 5.2. Equations (5.32) and (5.33) show that when $h_R = \overline{h}$, the first term in (5.33), $\mu_2(r, \varphi)$, is similar to (5.32). The first term in (5.33) can be associated with rare scatters, distributed far from the BS over a large area, as usually was postulated above for (5.10). Note that when $h_R > \overline{h}$, the second term in (5.33), $\mu_2(r, \phi)$, has a significant effect on the scatters distribution. The TOA–AOA distribution of the received multipath replicas at the BS, in a relative time-azimuth domain, was derived in [2, 15]. In these works, the TOA–AOA distributions in (5.32) and (5.33) were parameterized by the MS location $\mathbf{p}_j = (x_j, y_j) \,\forall j = 1, 2, \ldots, J$ in the x–y-plane as follows:

$$\widetilde{f}(t, \phi; \mathbf{p}_j) = \mu_1(t, \phi; \mathbf{p}_j) + \mu_2(t, \phi; \mathbf{p}_j)$$

$$= \frac{vd}{4} \left[\frac{h_R - \overline{h}}{\overline{h}} \widetilde{r} e^{-\gamma_0 \widetilde{r}} + \frac{\overline{h} r d \gamma_0}{h_R} \tau e^{-\gamma_0 r h/h} \right] \tag{5.34}$$

Note that when h_T, $h_R < \overline{h}$, that is, for 2-D case, expression (5.34) can be simplified as follows [2, 15]:

$$\widetilde{f}(t, \phi; \mathbf{p}_j) = 0.5v\gamma_0 \sin^2(\alpha/2)\tau d e^{-\gamma_0 \tau d} \tag{5.35}$$

where $\tau = tc/d$ is the time of arrival measured relatively to the pseudo-LOS, c is the speed of light, ϕ is the AOA of the scattered signal, and α is defined in Figure 5.2 as the grazing angle between the impinging and the reflected rays.

Considering now the straight-street waveguide scenario presented in Figures 5.4 and 5.5, the TOA–AOA distribution $\widetilde{f}(t, \pi; \mathbf{p}_j)$ parameterized by $\mathbf{p}_j$, which represents the existence of the wave mode induced by the multislit street waveguide at the distance of interest r from the MS, is [3, 4]

$$\widetilde{f}(t, \pi; \mathbf{p}_j) = \exp\left(-2\frac{|\ln \chi|}{a'(\phi)}\right) \tag{5.36}$$

where

$$a'(\phi) = a\sqrt{1 + \left(\frac{2a}{\lambda n}\right)^2}$$

[2, 15].

The joint model that combines the stochastic model, $\breve{f}(t, \pi; \mathbf{p}_j)$, from expression (5.34), which represents the random distribution of buildings surrounding the BS and the MS, and the multislit waveguide model, $\tilde{f}(t, \pi; \mathbf{p}_j)$, from (5.36), represents the scatterers that are aligned along the streets and induce the wave-guiding phenomena, describe different physical phenomena, and therefore, are independent. Figure 5.6 shows the scenario of interest, where the propagation conditions can be represented using the combination of these models, that is, by [2, 15]

$$f(t, \phi; \mathbf{p}_j) = \breve{f}(t, \pi; \mathbf{p}_j) \cdot \tilde{f}(t, \pi; \mathbf{p}_j) \tag{5.37}$$

Substituting the appropriate functions for $\breve{f}(t, \pi; \mathbf{p}_j)$ from (5.34) and (5.36), the distribution of all independent scatters as a function of AOA, ϕ, and TOA, τ, can be obtained as follows [2, 15]:

$$f(t, \pi; \mathbf{p}_j) = \frac{v\,d}{4}\left(\frac{h_R - h}{h}\tilde{r}e^{-\gamma_0\,\tilde{r}} + \frac{h\,r\,d\,\gamma_0}{h_R}\tau e^{-\gamma_0 rh/h_R}\right)e^{-2\frac{|\ln\chi|}{a'(\phi)}r} \tag{5.38}$$

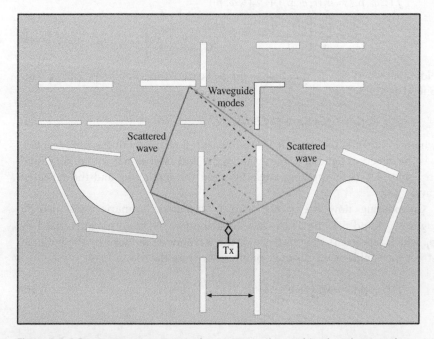

Figure 5.6 2-D representative scenario that motivates the combined stochastic and multislit waveguide model

Note that when $h_T < \bar{h}$ and $h_R < \bar{h}$, (5.38) can be rewritten as follows:

$$f(t, \phi; \mathbf{p}_j) = \frac{\gamma_0 v}{2} \sin^2\left(\frac{\alpha}{2}\right) \tau \, d e^{-\gamma_0 \tau d} e^{-2\frac{|\ln \chi|}{a'(\phi)} r} \tag{5.39}$$

where $a > \lambda$ and $a' = \frac{2a}{\lambda n}$.

5.2.2 Analysis of the Accuracy of MS Localization in Predefined Urban Scenarios

Here, the source localization performance of the maximum-likelihood (ML) and the minimum distance measure (MDM)-based methods [2, 15] is evaluated in various urban scenarios using ray-traced (Examples A, C–E) and collected (Example B) data. The distribution of the ray-traced data was estimated using a kernel method with the following bandwidths of the Gaussian window: $\sigma_1 = 50$ ns, $\sigma_2 = 1°$. The height of the MS antenna was considered to be above the ground level, $h_T = 1.5$ m, and remained constant in all scenarios.

Scenario I shown in Figure 5.4 was considered in Examples A and E, and Scenario II shown in Figure 5.5 was considered in Examples B–D. The performance of the proposed localization approach was evaluated using a root-mean-squared (RMS) localization error:

$$\text{RMS}(x_{MS}, y_{MS}) = \sqrt{(x_{MS} - \hat{x}_{MS})^2 + (y_{MS} - \hat{y}_{MS})^2} \tag{5.40}$$

the distribution of the RMS localization error in the TOA–AOA domain, and its cumulative distribution functions (CDFs).

5.2.2.1 Example 1: The statistical model vs. ray-tracing simulation according to the topographic map

In this Example, the fit between the statistically modeled and the ray-traced TOA–AOA distributions in Scenario I (from Figure 5.4) was analyzed. Figure 5.7 shows the numerically estimated and statistically predicted distributions of the TOA–AOA measurements by use of stochastic (a) and ray-tracing (b) approaches.

One can notice the similarity between these distributions. Additionally, in [15] it was shown that the statistical model accurately represents the collected TOA–AOA measurements, and this example demonstrates that it also accurately represents the measurements generated by ray tracing.

5.2.2.2 Example 2: MS and BS antennas are below the rooftop level

In this example, the source localization performance of the proposed methods was evaluated using collected data from Scenario II in Figure 5.5 with the following parameters: $\gamma_0 = 4$, $v = 100$, $\lambda = 0.16$, and $\chi = 0.5$. The BS antenna was located below the rooftop, at $h_R = 10$ m, and the MS was located at $\phi = 23°$, $d = 0.3$ km. Figure 5.8a–d shows the distributions of the RMS localization error in (5.40) using different probabilistic methods: MDM-Kendel likelihood

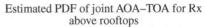

Estimated PDF of joint AOA–TOA for Rx
above rooftops

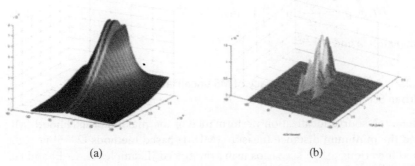

(a)   (b)

Figure 5.7 Example 1: TOA–AOA distribution: (a) predicted by the statistical model and (b) estimated from the generated receiving personal subscriber (RPS) measurements and the corresponding ray tracing

distance (KLD), MDM-Hellinger, MDM-inter-symbol error (ISE), and ML [2, 15], respectively.

In Figures 5.8a-d by higher brightness curvers, can be interpreted as follows: the lower the probability of the RMS localization error at some location in the TOA–AOA domain, the higher the probability that scatterers at this location were illuminated by the MS. The multimodal behavior of probabilities of RMS localization error can be justified by the fact that in non-line-of-sight (NLOS) propagation condition, only a subset of scatterers that is illuminated by the MS is observable, and therefore, only locations of the scatterers in this subset can be estimated. We use the statistical model of the propagation conditions in the urban environment to recognize the MS location that illuminates some particular subset of scatterers, which induces the TOA–AOA distribution at the BS. When the MS and the BS antennas are below the rooftop level, the MS illuminates both the local and distant scatterers (via multislit waveguides), whose reflections are received at the BS. As a result, multiple subsets of scatterers are illuminated, inducing the multimodal behavior of the RMS localization error in the TOA–AOA domain.

It should be noticed that the RMS localization errors of 54, 57, 97, and 131 m were obtained for the MDM-KLD, MDM-Hellinger distance, MDM-ISE, and ML methods, respectively. Note that the MDM-based method with all tested distance measures outperforms the ML method and that the MDM-KLD outperforms other tested methods. Note that the superiority of the MDM-based method stems from its robustness to the modeling mismatch in scenarios where the propagation conditions cannot be fully represented by the proposed statistical model. The superiority of the MDM-KLD method can be explained by the fact that the KLD emphasizes the influence of the local scatterers that surround the MS.

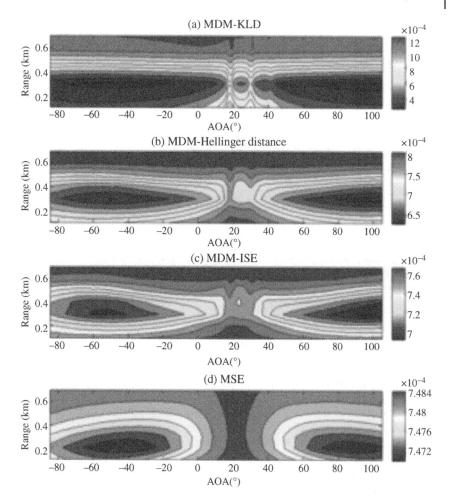

Figure 5.8 Example 2: Probability of RMS localization error when MS and BS antennas are below the rooftop level, by use of several probabilistic methods: (a) MDM-KLD, (b) MDM-Hellinger distance, (c) MDM-ISE, and (d) ML (according to [15])

5.2.2.3 Example 3: MS antenna is below and BS antenna is above the rooftop level

Considering the Scenario II in Figure 5.5 with the BS antenna above the rooftop level, $h_r = 45$ m, the influence of the BS antenna height on the source localization performance is analyzed in this example. Figure 5.9a–c shows the 2-D distributions of the probability of RMS localization error of the MDM-KLD, the MDM-Hellinger distance, and the MDM-ISE methods, respectively.

One can notice the improvement in the probability of RMS localization error, when compared with the results in Figure 5.9. This improvement is associated with the fact that the increased height of the BS antenna minimizes

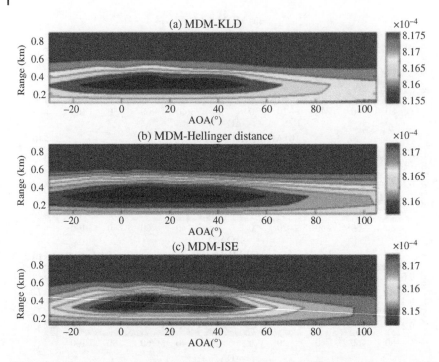

Figure 5.9 Example 3: Probability of RMS localization error when BS antennas is above the rooftop level: (a) MDM-KLD, (b) MDM-Hellinger distance, and (c) MDM-ISE [4, 5]

the illumination of distant scatterers and therefore diminishes their influence. The localized illumination of the scatterers surrounding the MS induces a more concentrated TOA–AOA distribution at the BS via the enhanced over-the-rooftop diffraction. In this example, the RMS localization errors were obtained for the MDM-KLD, MDM-Hellinger, and MDM-ISE methods, respectively. This example demonstrates the localization accuracy of the proposed approach in the AOA domain. Consider the BS antenna below the rooftop level and the urban environment in Scenario II in Figure 5.9 in polar coordinates with the BS at its origin. In this example, the location of the MS was shifted by anticlockwise relative to the MS location in Example 2. Figure 5.10a–c shows the 2-D distributions of the probability of RMS localization error of the MDM-KLD, the MDM-Hellinger distance, and the MDM-ISE methods, respectively.

Comparing Figures 5.8 and 5.10, notice similar localization errors and the shift in the distribution of the RMS localization error, which matches the shift in the MS location.

5.2.2.4 Example 4: Multiple MS locations
The localization performance of the proposed approach for multiple MS locations and BS antenna heights was evaluated in this example using the CDF

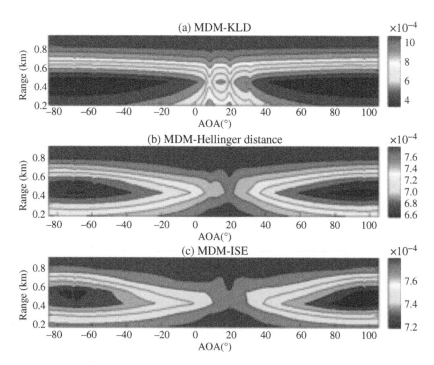

Figure 5.10 Example D: Probability of RMS localization error when MS and BS antennas are below the rooftop level, and the location of the MS is shifted by(a) MDM-KLD, (b) MDM-Hellinger distance, and (c) MDM-ISE [2, 15]

of the RMS localization error. The multiuser environment with 18 homogeneously selected MS locations in scenario I in Figure 5.4 was considered. First, the area of interest was discretized in polar coordinates into J potential MS locations with resolution of 0.1 μm by 1°. Next, the statistical models of the TOA–AOA distributions for each hypothesized MS location were obtained. Three heights of the BS antenna above, $h_R = 30$, 45, 60 m, and one height below rooftop level, $h_R = 10$ m, were simulated in Example 4. Figure 5.11 shows the CDF of the RMS localization error of the MDM-based method with KLD, Hellinger distance, and the ISE measures.

In Figure 5.11, any value in the y-axis represents the probability that the RMS localization error is smaller than the corresponding value in the x-axis. From the results illustrated by Figure 5.11, one can obtain that more than 80% of MSs can be localized with the RMS location error of 200 m and with RMS location error of 50 m. Note that the performance of MDM-based method with all distance measures is similar.

Figure 5.12 shows the influence of the BS antenna height on the localization accuracy. The performance of the MDM-KLD method was evaluated in this example. Notice the 50% improvement in the RMS localization error for the

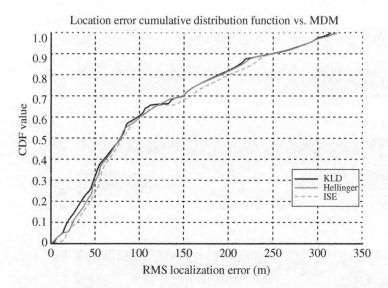

Figure 5.11 Example E: CDF of RMS localization error of the MDM-KLD, MDM-Hellinger distance, and MDM-ISE methods [15]

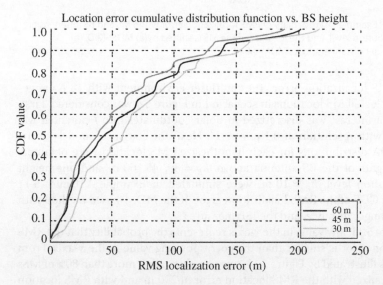

Figure 5.12 Example E: CDF of RMS localization error of the MDM-KLD method when BS antenna heights are $h_R = [30, 45, 60]$ m according to [15]

BS antenna at height of $h_R = 60$ m, when calculated for the RMS localization error.

Comparison between Figures 5.11 and 5.12 demonstrates an improvement of about 100% in the RMS localization error when calculated for the 50 m RMS location error.

Finally, Figure 5.13 shows the received signal distributions, induced by(a) random scattering, (b) one waveguide "response,"(c) two waveguides "response," and (d) local scattering in the MS proximity in the above-the-rooftop propagation scenario.

All these scenarios are shown schematically in Figure 5.14.

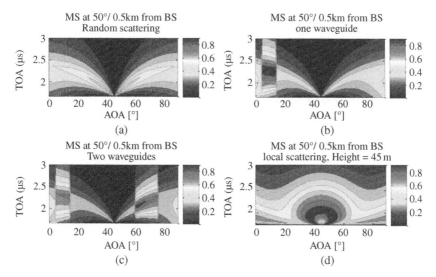

Figure 5.13 Received signal distributions in the TOA–AOA domain induced by (a) the random scattering, (b) one waveguide, (c) two waveguides, and (d) local scattering in the above-the-rooftop propagation scenario

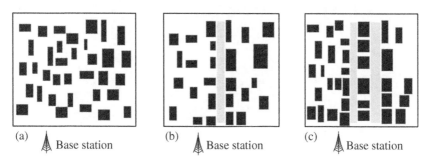

Figure 5.14 Three scenarios of buildings' layout: (a) randomly distributed buildings, (b) existence of one street-waveguide, and (c) existence of two street-waveguides. The street-waveguides are marked in gray

It can be outlined that Figures 5.8, 5.10, and 5.13 support our conjecture that the MS transmission from different geographical locations induces significantly different received signal distributions in the TOA–AOA domain that could be exploited for the MS localization.

Next, we present some practical aspects obtained from the proposed statistical approach. Using now the probabilistic approach based on maximum likelihood level (MLL)model (e.g. on likelihood function maximization algorithm proposed in [15]), and by use of the above formulas (5.37)–(5.40), we present in Figure 5.15 the AOA–TOA likelihood function described distribution of signal power in a joint azimuth-time-delay 2-D domain.

One can notice the multimodal nature of the 2-D likelihood function. This multimodality can be explained by the fact that when the MS and BS antennas are below the rooftop level, the MS illuminates both the local and distant scatterers (via the multislit waveguides), whose reflections are received at the BS. As a result, multiple subsets of scatterers are illuminated, inducing multimodal behavior of the likelihood function in the AOA–TOA domain. Note that this multimodality is a natural behavior of the ML decision rule (i.e. maximum probability to localize a subscriber) in case of NLOS propagation conditions due to the fact that this ML result shows the distribution of scatterers that

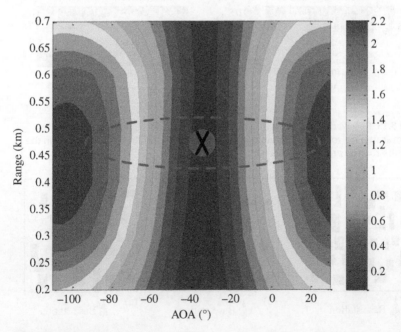

Figure 5.15 AOA–TOA likelihood functions, when BS is below the rooftop level. The true MS position marked by a cross is at AOA = −41° (the estimated AOA is −37°) and at a distance (true and estimated) of 420 m from the BS

have maximum likelihood to be illuminated by MS and those scatterers are distributed in space domain, surrounding the MS.

Moreover, according to [2, 15], the accuracy of MS location in the scenario I, where BS antenna is lower than the building's roofs (see Figure 5.4), was estimated as $\phi = -37°$ and $d = 0.4$ km. Comparing this result to the true MS location at $\phi = -41°$ and $d = 0.4$ km, the localization performance of the proposed statistical approach is evaluated using the RMS localization error described by Eq. (5.40). For the considered scenario shown in Figures 5.4 and 5.5, the localization RMSE of 32 m was achieved. Note that here the simulated model is derived for a straight-street crossing configuration, and different street structures are expected to generate different spreading phenomena.

Next, a scenario corresponding to that shown in Figures 5.4 and 5.5, where the BS antenna is placed on the rooftop level, but MS antenna – below the rooftops level, was analyzed in [2, 15]. Figure 5.16 shows the likelihood function of the AOA–TOA measurements.

The localization mean square error in the distance domain, RMSE = 48 m, was achieved in this scenario, that is the accuracy in this scenario is approximately less than twice than that in the case of the first scenario. An additional scenario is shown in Figures 5.4 and 5.5 for the second scenario, where the BS

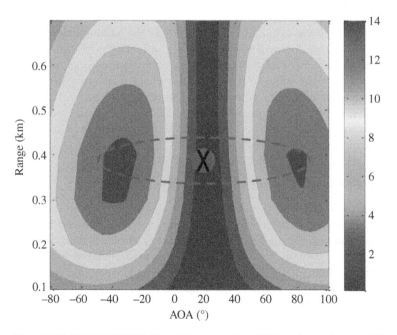

Figure 5.16 ML AOA–TOA likelihood functions, when BS is on the rooftop level. The true MS position marked by a cross is at AOA = −23° and at a distance of 310 m from the BS. The RMSE = 48 m

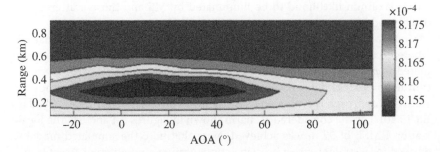

Figure 5.17 The AOA–TOA likelihood functions, when BS is above the rooftop level. The true MS position is around AOA = 22°, a distance of ~ 300 m (true and estimated) from the BS

antenna is placed above the rooftops, and the MS antenna below the rooftops. This scenario was simulated by Refs. [2, 15] and is presented in Figure 5.17.

The true MS position was considered to be at AOA = 22° and at a distance of 300 m from the BS.

5.3 Positioning of Any Subscriber in Multiuser Land–Atmosphere Communication Links

The situation with effects of buildings' overlay profile and buildings' density layout on any flying subscriber will not be changed considering now land–aircraft communication links, because the same features of built-up topographic map are influenced on positioning of flying and stationary or moving subscriber located at the areas of service. The scenario, sketched in Figure 5.18, allows us to analyze signal power distribution in AOA, TOA, and DS domains for predefined built-up area – local airport placed in suburban area; the sketch of the topographic map is presented in Figure 5.19.

5.3.1 Signal Distribution in the Time-Delay Domain

We start with evaluation of the time delay for each from five aircrafts with respect to the ground-based receiver of the first operator, denoted by Rx-1 based on formulas (5.16)–(5.19). For this operator, the total time delay (in microsecond) is shown in Figure 5.20a, and its standard deviation (called spreading) is shown in Figure 5.20b.

Similar total time-delay distribution and its spread for the operator Rx-2 and Rx-3 are shown in Figures 5.21 and 5.22, respectively.

The results shown in Figures 5.20a–5.22 allow us to predict apriori the total time delay and its standard deviation from each aircraft with respect to each other, compared to the ground-based receivers–operators. The complicated distribution of the total and relative (to each transmitter with respect to each

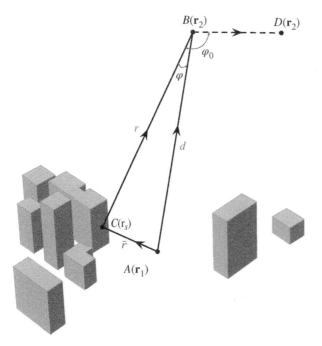

Figure 5.18 Geometry of buildings' effects on land–aircraft communication

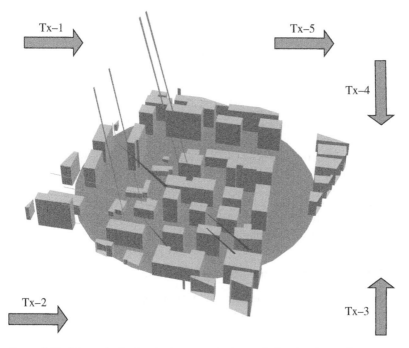

Figure 5.19 Schematically sketched communication with the virtual aircrafts, as transmitters, denoted by Tx-1, Tx-2, Tx-3, Tx-4, and Tx-5, with their own evaluation and direction with respect to the receiving operator antenna

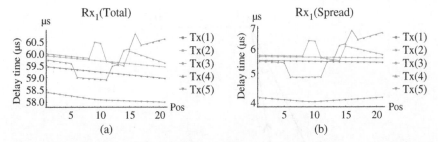

Figure 5.20 (a) The total delay and (b) the delay spread of each from five transmitters located at different aircrafts with respect to the first receiver Rx-1

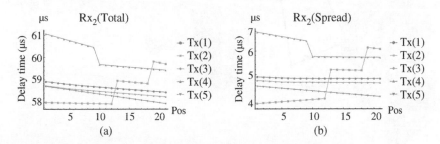

Figure 5.21 (a) The total delay and (b) the delay spread of each from five transmitters located at different aircrafts with respect to the first receiver Rx-2

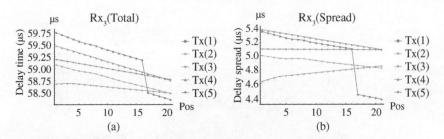

Figure 5.22 (a) The total delay and (b) the delay spread of each from five transmitters located at different aircrafts with respect to the first receiver Rx-3

receiver) time delay and delay spread, which obviously depend on angle of elevation of each aircraft antenna, its position, and height orientation with respect to each operator (Rx) antenna, is clearly seen.

5.3.2 Signal Distribution in the Doppler-Shift Domain

Previously, two models, 2Dmodel without accounting of the buildings' profile and which is correct only for land-to-land links with law-elevated BS and

Deviation from doppler shift with LOS conditions

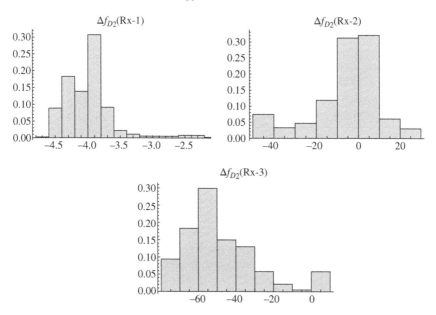

Figure 5.23 Deviation from Doppler shift in LOS conditions for three ground-based antennas, Rx-1 to Rx-3

MS antennas and then 3Dmodel, described by (5.31), were checked. Because the 3Dmodel, according to formula (5.31), is a more realistic approach for land–aircraft communication links, taking into account both the buildings' overlay profile and the buildings' density, we present here only the results of computations of this model for the same scenario of existence of three operator antennas located at points Rx[1], Rx[2], and Rx[3]. The corresponding results of numerical computations are shown in Figure 5.23 for all three operator antennas. Here, the cumulative effects of deviations of Doppler shift (in Hertz) for the five subscribers, Tx-1 to Tx-5, are presented in the form of histogram for each orientation and elevation of the corresponding Tx and Rx antennas.

The cumulative effects of the means and standard deviations (called Doppler spread) of Doppler frequency (f_D) and its shift (Δf_D) are presented in Table 5.1, according to computations made based on formula (5.31), for three operator antennas: top row – for Rx-1, middle row – for Rx-2, and bottom row – for Rx-3.

It is clearly seen that depending on the orientation of aircrafts with respect to the operator antennas, the mean Doppler frequency and its deviation can be as negative and positive but do not exceed several tens of Hertz and several Hertz, respectively.

Table 5.1 Mean Doppler frequency, its STD and mean shift (Δf_D) and its STD for three position of Rx antennas: Rx-1, Rx-2, and Rx-3

Mean (f_D)	STD (f_D)	Shift (Δf_D)	STD (Δf_D)
61.5264	1.25483	−3.56047	1.26074
−37.8394	16.97	−5.94446	15.9525
54.5572	20.009	−48.0621	19.4402

References

1 Blaunstein, N., Toeltsch, M., Christodoulou, C. et al. (2002). Azimuth, elevation and time delay distribution in urban wireless communication channels. *J. Anten. Propag. Mag.* 48(3): 425–434.

2 Tsalolihin, E., Bilik, I., and Blaunstein, N. (2011). A single-base-station localization approach using a statistical model of the NLOS propagation conditions in urban terrain. *IEEE Trans. Veh. Technol.* 60 (3): 1124–1137.

3 Blaunstein, N. and Levin, M. (1996). VHF/UHF wave attenuation in a city with regularly spaced buildings. *Radio Sci.* 31 (2): 313–323.

4 Blaunstein, N. (1998). Average field attenuation in the non-regular impedance street waveguide. *IEEE Trans. Anten. Propag.* 46 (12): 1782–1789.

5 Blaunstein, N. and Christodoulou, C. (2007). *Radio Propagation and Adaptive Antennas for Wireless Communication Links*, 1e. Hoboken, New Jersey: Wiley.

6 Blaunstein, N. and Christodoulou, C. (2014). *Radio Propagation and Adaptive Antennas for Wireless Communication Networks – Terrestrial, Atmospheric and Ionospheric*, 2e. Hoboken, New Jersey: Wiley.

7 Blaunstein, N. (2000). Distribution of angle–of–arrival and delay from array of building placed on rough terrain for various elevation of base station antenna. *J. Commun. Netw.* 2 (4): 305–316.

8 Hayakawa, M., Katz, D., and Blaunstein, N. (2008). Signal power distribution in time delay in Tokyo City experimental sites. *Radio Sci.* 43, RS3006: 1–9.

9 Blaunstein, N., Yarkony, N., and Katz, D. (2006). Spatial and temporal distribution of the VHF/UHF radio waves in built-up land communication links. *IEEE Trans. Anten. Propag.* 54 (8): 2345–2356.

10 Blaunstein, N. and Tsalolihin, E. (2004). Signal distribution in the azimuth, elevation and time delay domains in urban radio communication links. *IEEE Anten. Propag. Mag.* 46 (5): 101–109.

11 Blaunstein, N., Toulch, M., Laurila, J. et al. (2006). Signal power distribution in the azimuth, elevation and time delay domains in urban

environments for various elevations of base station antenna. *IEEE Trans. Anten. Propag.* 54 (10): 2902–2916.

12 Blaunstein, N. and Ben-Shimol, Y. (2004). Frequency dependence of path loss characteristics and link budget design for various terrestrial communication links. *IEEE Trans. Anten. Propag.* 52 (10): 2719–2729.

13 Blaunstein, N. and Ben-Shimol, Y. (2006). Spectral properties of signal fading and Doppler spectra distribution in urban communication mobile links. *Wireless Commun. Mob. Comput.* 6 (1): 113–126.

14 Blaunstein, N., Katz, D., and Hayakawa, M. (2010). Spectral properties of modulated signal in the Doppler domain in urban multipath radio channels with fading. *IEEE Trans. Anten. Propag.* 58 (8): 2795–2800.

15 Tsalolihin, E., Bilik, I., and Blaunstein, N. (2010). Mobile user location in dense urban environment using unified statistical model. *European Conference on Antennas and Propagation* (EuCAP 2010), Barcelona, Spain (12–16 April 2010), pp. 157–158.

Part III

Advanced Integrated-Cell Technologies for Modern 4G and 5G Networks

6

Femto/Pico/Micro/Macrocell Network Deployments for Fourth and Fifth Generations

The femtocell concept, introduced recently in practice of wireless cellular and noncellular communication networks design, is one of the examples of new technologies that makes use of the existing 3-G homogeneous and nonhomogeneous cellular networks to yield high-speed mobile communications (up to 1 Gbit/s) [1–12]. A femtocell concept was introduced as an example of how to improve grade-of-service (GoS) and quality-of-service (QoS) for modern multiple-input-multiple-output (MIMO) networks by introducing a nonstandard planning of the cells' pattern instead of the existing technologies of the cells' pattern design [13–16].

According to Refs. [1–12], we define femtocell as the *home access points* (HAPs) (also called *femto-APs*) arranged inside the existing micro- or macrocell networks for the increasing of the rate of information data stream for each home subscriber. The increasing demand observed during the recent decades for higher information data rate in standard wireless networks has triggered the performance of advanced cellular technologies and modern networks. Thus, the third-generation partnership projects (3PPP and 3PP2) [4, 5], high-speed packet access (HSPA) [11], wideband code-division multiple access (WCDMA-2000) and WCDMA/HSPA standards, and the modern fourth-generation (4G) technologies and the corresponding networks on the basis of WiMAX (based on the protocol 802.16e) and LTE standards [1, 3, 8] were continuously adopted to supply the real mobile broadband experience for the mobile customers. This occurs because of the growing demand for mobile wireless communications, which requires to determine and to fully understand the capacity limitations of each technology due to the fact that the capacity limit formulates the maximum data rates based on the channel capacity equation that was introduced firstly by Shannon [2].

One of the main approaches to overcome the fundamental capacity limitations and as a result to increase the maximum user throughput is to use the higher order modulation techniques for broadband networks, such as orthogonal frequency division multiplexing (OFDM) [9]. However, such a technique requires smaller cell radius, especially when the reuse of one frequency scheme is used.

Advanced Technologies and Wireless Networks Beyond 4G, First Edition.
Nathan Blaunstein and Yehuda Ben-Shimol.
© 2021 John Wiley & Sons, Inc. Published 2021 by John Wiley & Sons, Inc.

Femtocell access point (FAP) networks have recently received considerable attention from communication society due to their enormous potential for capacity improvements by answering the small cell radius requirement [1]. The initial standardization of FAP by the 3GPP, 3GPP2, and WiMAX Forum, [3–5, 8, 12] was completed lately, signaling that femtocell technology has been recognized by the highest profile standardization structures worldwide.

Usually, femtocells are installed for short-range, low-cost, and low-power APs for better voice and information data reception, mostly for indoor wire or wireless communication links. Therefore, the main benefits of FAP networks include improvements in indoor radio coverage with a significant decrease in the transmitting power, becoming noninvasive for each user, with longer mobile subscriber (MS) battery life (due to shorter distance of users' service), with additional value-added services related to the location of FAPs in the proximity of the MS household and, of course, the capital expense (CAPEX) saving in the macrocell network infrastructure and core elements. Finally, the femtocell concept allows designers of such combined femtocell–microcell or femtocell–macrocell networks to achieve a higher capacity, that is, maximum possible rate of signal data passing via such indoor–outdoor channels with minimization in the multiplicative noise due to fading and the noise caused by interference between users [1, 10–12]. At the same time, we notice that the coexistence between the FAP, neighboring FAPs, and micro- or macrocell BSs (denoted as MBS) remains a key problem that needs to be addressed [3–5].

A precise control of femto–macrocellular (FMC) interference should be performed. This control relates to subband scheduling and interference cancellation. Thus, to control FMC interference, the macrocell bandwidth should be split into subbands, and the short-range femtocell links should allow their power across these subbands. The corresponding procedure of the macrocell partitioning into subbands is done as follows.

The subband Δf_i, $i \in [1, N]$, is used in all M cells and serves users located close to each cell's BS. The other subbands Δf_j, $j \neq i$, serve users close to the cells edges. During subband partitioning, an adaptive power control technique is usually used. Such approach of powers allocation across the subbands can maximize loading and GOS.

Finally, using femtocell APs, we can satisfy the problem of macrocell interface, reduce intercell interference between adjacent macrocells, mitigate the femtocell–macrocell user's interference, and improve cell edge users' performance. The main goal of femtocell–macrocell and femtocell–microcell deployments is to avoid cochannel interference and to increase the overall cell capacity, which in turn allows achieving high data rate for each indoor subscriber.

All these benefits depend strongly on the propagation phenomena (i.e. physical background) that occur in indoor–outdoor communication environments. A good prediction of these propagation effects allows solving the problem of full radio coverage for each femtocell (e.g. indoor) subscriber located in the area of each FAPs service.

Here, we focus our effort on the coexistence analysis that takes into consideration the different scenarios of FAP deployment. The analysis is performed in terms of channel capacity estimation for different FAP deployment strategies. Following the [8, 9], we formulate the capacity analysis on the basis of four main terminologies:

- Dedicated spectral assignment (DSA), that is, the FAP deployment using a dedicated spectrum that is not used for the macrocell network;
- Shared spectral assignment (SSA), that is, the FAP deployment using the same frequency carrier as a macrocell network;
- Closed subscriber group (CSG), that is, the FAP is accessible only for a local group of users according to the defined access list;
- Open subscriber group (OSG), that is, all MBSs might access the FAP coverage service.

We analyze all the combinations, DSA with CSG and OSG, as well as SSA with CSG and OSG, below. We also notice that in the CSG case, the only users, which are inside the indoor environment, are assumed to be served by the FAP. Some of the details on shared spectral channel assignment deployment can be found in [8–10]; the comparison between the closed and open access group lists is analyzed in [11], and different power control techniques are discussed in [10]. The femto forum white papers have also included studies for coexistence analysis in terms of mutual interference [12].

6.1 Channel Capacity Models in Integrated Femtocell–Microcell/Macrocell Networks

To analyze the potential of channel capacity of mobile users in networks with integrated femtocell deployments, Shannon's equations were introduced considering different FAP available configurations (CSG and OSG) and spectral channel assignment strategies (DSA and SSA) mentioned above. The following conditions in such femtocell deployments and channel capacity models were assumed:

(a) The total spectral bandwidth of the system, B_t, is assigned to FAPs and MBSs according to the considered configuration – DSA, SSA, CSG, or OSG.
(b) All users have the same available bandwidth, which is assigned to comply with the highest available demand of service in the network, that is, each MS receives the equivalent part of the B_t.
(c) All FAPs and MBSs transmit simultaneously to all active subscribers, whether stationary or mobile.

We notice here that to calculate the average received signal strength (RSS) accounting for slow- and fast-fading phenomena caused by multiple scattering, diffraction, and reflection, the unified propagation models for outdoor, indoor, and outdoor–indoor scenarios were used following Refs. [13–22]. Now, let us introduce the main formulas for considering different FAP available configurations (CSG and OSG) and spectral channel assignment strategies (DSA and SSA).

6.1.1 Shared Spectrum Assignment (SSA) with Closed Subscriber Group (CSG)

In the case when the total spectrum bandwidth is shared between the FAP network and the macrocell BSs (MBSs), the capacity of MS user i_F in FAP coverage can be introduced as follows [15, 16]:

$$
C_{\text{SSA\_CSG\_indoor\_}i} = B_{tN} \log_2 \left(1 + \frac{S_{Fi}}{kT B_{tN} + \sum_{j=1}^{J} I_j + \sum_{\substack{l=1 \\ l \neq i}}^{L} I_l} \right) \tag{6.1}
$$

Here, $B_{tN} = B_t/N$ is a bandwidth normalized to the number of users N served by FAP; S_{Fi} is the signal strength of the FAP at the location of MS user i_F served by FAP with $i_F \in \{1, 2, \ldots, N\}$; I_l is the interference strength of the MBS antenna with $l \in \{1, 2, \ldots, L\}$, where L is the total number of MBSs; I_j is the interference strength of the neighboring FAP j, $j \in \{1, 2, \ldots, J\}$, where J is the total number of FAPs; and $k_B T B_t$ is the thermal noise, where k_B is the Boltzmann coefficient and T is the temperature (in Kelvin). We notice that the MS users, which were allocated indoor, are considered to be registered in the CSG; otherwise, they are not allowed to be served by FAP. The capacity of the outdoor MS user i_F can be calculated as follows [15, 16]:

$$
C_{\text{SSA\_CSG\_outdoor\_}i} = B_{tP} \log_2 \left(1 + \frac{S_{Mi}}{k T B_{tP} + \sum_{j=1}^{J} I_j + \sum_{\substack{l=1 \\ l \neq i}}^{L} I_l} \right) \tag{6.2}
$$

where $B_t P$ is a bandwidth normalized to the number of users P served by an MS, and $S_M i$ is a signal strength of the MBS at the location of MS user i_M served by MBS with $i_M \in \{1, 2, \ldots, P\}$.

6.1.2 Shared Spectrum Assignment (SSA) with (OSG)

The main difference for this case is that the MS users that are located outdoor can be potentially served by FAP. Therefore, for MS users i_{F_o}, which are served by FAP and located outdoor, the channel capacity can be written as [15, 16]

$$C_{\text{SSA\_CSG\_outdoor\_}iF} = B_{tN}\log_2\left(1 + \frac{S_{Fi}}{k\,T\,B_{tN} + \sum_{j=1}^{J}I_j + \sum_{\substack{l=1 \\ l\neq i}}^{L}I_l}\right) \tag{6.3}$$

The decision of the outdoor MS served by FAP or MBS antenna is done by simple handover threshold; that is, the transmitter (FAP or MBS antenna) with highest signal strength gets to serve the MS.

6.1.3 Dedicated Spectrum Assignment (DSA) with Closed Subscriber Group (CSG)

In case the dedicated spectrum is assigned to the femtocell network, there is no mutual interference between the MBS and the FAP; however, the total bandwidth B_t is divided between the FAP and MBS networks in some manner, that is, the allocation can be potentially not symmetrical. Therefore, the channel capacity for the MS user iF that is in FAP coverage can be introduced as follows [15, 16]:

$$C_{\text{DSA\_CSG\_indoor\_}i} = B_{tNd}\log_2\left(1 + \frac{S_{Fi}}{k\,T\,B_{tNd} + \sum_{\substack{j=1 \\ j\neq i}}^{J}I_j}\right) \tag{6.4}$$

where $B_{tNd} = B_t\text{FNR}/N$, and FNR is a FAP network ratio, which defines the part of the total B_t spectrum allocated for the FAP network. The capacity of the outdoor MS user i_M, which is served by MBS, can be calculated as follows [15, 16]:

$$C_{\text{DSA\_CSG\_outoor\_}i} = B_{tNd}\log_2\left(\frac{S_{Fi}}{B_{tNd} + \sum_{j=1}^{J}I_j}\right) \tag{6.5}$$

where $B_{tPd} = B_t\text{FNR}/P$ is, as above, a bandwidth normalized to the number of users P served by the MBS.

6.1.4 Dedicated spectrum assignment (DSA) with open subscriber group (OSG)

In such a scenario, the channel capacity for the MS users, i_F, that is, located under the FAP radio coverage (having better signal strength), can be introduced as follows [15, 16]:

$$C_{\text{DSA\_OSG\_outoor\_}i} = B_{tNd}\log_2\left(\frac{S_{Fi}}{B_{tNd} + \sum_{j=1}^{J}I_j}\right) \tag{6.6}$$

where $B_{tNd} = N_t\text{FNR}/N$ is, as above, a bandwidth normalized to the number of users N served by the MS antenna.

Addressing the problem of optimal resources allocation in the predefined built-up areas of interest and for the above four scenarios, let us introduce a well-known power allocation procedure called in the literature the "spatial water filling" [15, 16].

6.2 Analysis of Femto/Pico/Micro/Macrocell Networks Based on Propagation Phenomena

In our numerical analysis of the four scenarios described above, we assume that the users are randomly, but uniformly, distributed in the investigated area of service. For the analysis, two representative areas were selected: one is for an urban area and the other is for a suburban area (see definitions in [13, 14]).

We also assume that each FAP has three categories corresponded to the output antenna power of 10, 15, 21 dBm, respectively. In each category, the power dynamic range is 30 dB. Each FAP affects other users within a radius of 150 m.

For each distribution described above, we need to optimize the maximum transmitted power. The optimization criterion is to maximize the site's total ergodic capacity, that is, the sum of maximum available capacity for each user in the site under investigation [15, 16]:

$$C_{\text{total}} = \sum_i C_i(\text{outdoor users}) + \sum_i C_i(\text{outdoor users without femto}) +$$

$$\sum_i C_i(\text{outdoor users with femto}) \qquad (6.7)$$

6.2.1 Propagation Aspects in Integrated Indoor and Outdoor Communication Links

To analyze various scenarios in the outdoor–indoor communication environments for the femtocell–macrocell joint planning tool deployment, the multiparametric stochastic approach for signal strength prediction in the urban environment is used. This approach is fully described in [13, 14], and some details on the stochastic models for regular and nonregular distributions of buildings in the urban environment are presented in [13–22]. This stochastic approach combines the multipath propagation along straight crossing streets, areas surrounding the streets, and other natural or man-made obstructions randomly distributed (according to *Poisson's* law as an ordinary flow of scatterers) on the rough terrain. General formulas were obtained in Refs. [17–20] for prediction of the signal path loss in various scenarios with different elevations of the base station and the subscriber antennas. In this paragraph, the simplified approach based on [19] is proposed, where multiple diffraction and scattering, having the coherent and incoherent parts, have been rearranged in the simple forms.

6.2.1.1 Outdoor propagation model

Many propagation models have been developed in recent two to three decades for investigation of outdoor propagation channels in different communication environments: rural, suburban, mixed residential, and urban with various configurations and densities of obstructions (homes, buildings, hills, trees, vegetable, etc.) [13, 14, 17–27]. For the investigation of the response of outdoor communication channels for the practical applications to macrocell/femtocell networks deployment, the multiparametric stochastic approach for signal strength prediction in the urban environment was introduced recently in the previous works [17–27]. The approach, summarized in [13, 14], combines the multipath propagation along straight crossing streets, areas surrounding the streets, and other natural or man-made obstructions randomly distributed (according to Poisson law as a ordinary flow of scatterers) on the rough terrain. General formulas were obtained for prediction of the signal path loss in various scenarios with different elevations of the base station and the subscriber antennas. Below, we follow the simplified approach introduced in [14], where multiple diffraction and scattering, having the coherent and incoherent parts, have been rearranged in the simple forms of "straight-line" equations, as it is usually proposed by other authors [17, 18, 21, 22]. Depending on the elevation of the BS antenna with respect to the building rooftops and the MS antenna, there are several scenarios occurring in the built-up scene that were proposed in [14] according to [13]. Here, we introduce the most general scenario, which is really observed in the urban and suburban environments:

(a) when the BS antennas are located above the rooftop level and
(b) on the rooftop level, but the mobile or stationary subscriber is located below the rooftop level in both cases, and multiple scattering and single or multiple over-rooftop diffraction occurred (see [13, 14]).

In the first scenario, as shown in Figure 6.1, in non-line-of-sight (NLOS) conditions, diffraction from roofs, located close to the MS antenna, is the source

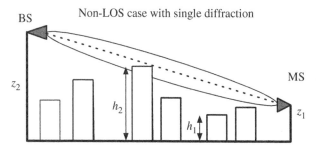

Figure 6.1 First scenario: BS is above and MS is below the rooftops

of shadowing and the slow-fading phenomenon. Here, following [14] (see also Chapter 3 where the stochastic model is presented), we get

$$L(r) = -32.4 - 30 \log_{10} f_{(MHz)} - 30 \log_{10} r_{(km)} - L_{fading} \tag{6.8}$$

where in this specific case

$$L_{fading} = 10 \log_{10} \frac{\gamma_0 \ell_v F^2(z_1, z_2)}{|\Gamma| \sqrt{\frac{\lambda r}{4\pi^3} + (z_2 - \overline{h})^2}} \tag{6.9}$$

Here, r is the range between the terminal antennas in km; z_1 and z_2 are the heights of the MS and BS antennas, respectively; $\gamma_0 = 2\overline{L}v/\pi$ is the density of building contours in km^{-1}; $\overline{L}$ is the average length of the buildings in km; and v is the number of buildings in one square kilometer. Here also, the buildings' overlay profile is simply $F^2(z_1, z_2) = (\overline{h} - z_1)^2$ for the case of uniformly distributed building height from $h_1 \equiv h_{min}$ to $h_2 \equiv h_{max}$, and $\overline{h} = (h_{min} + h_{max})/2$; ℓ_v is the average wall's roughness parameter [13, 14], usually equals 1–3 m; and $|\Gamma|$ is the absolute value of reflection coefficient (see [25]):

$$|\Gamma| = \begin{cases} 0.45 & \text{glass} \\ 0.5 - 0.6 & \text{wood} \\ 0.7 - 0.8 & \text{stones} \\ 0.9 & \text{concrete} \end{cases}.$$

The wavelength of radio wave in cellular communication spans over a wide range from $\lambda = 0.05$ to 0.53 m, covering most of the modern wireless networks. Equation (6.9) can be used for link budget design for various microcell–macrocell scenarios occurring in the built-up terrain and for high-elevated BS antenna with respect to the average profile of the buildings.

In the case of the second scenario, depicted in Figure 6.2, diffractions from roofs of the buildings, located close to the MS and BS antennas, are the sources of shadowing and the fading phenomena. Here, we get [14]:

$$L(r) = -41.3 - 30 \log_{10} f_{(MHz)} - 30 \log_{10} r_{(km)} - L_{fading} \tag{6.10}$$

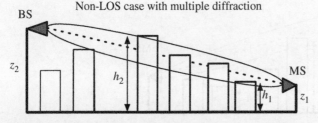

Figure 6.2 Second scenario: BS is at the same level or lower than h_2, and MS is lower than h_1

where

$$L_{\text{fading}} = 10 \log_{10} \frac{\gamma_0^4 \ell_v^3 F^4(z_1, z_2)}{\lambda |\Gamma|^2 \left[\frac{\lambda r}{4\pi^3} + (z_2 - \overline{h})^2 \right]} \tag{6.11}$$

All the parameters presented in (6.10) and (6.11) are the same, as introduced in (6.8) and (6.9). In this, a more complicated multiray microcell–macrocell scenario, the buildings' overlay profile function can be presented as [14]

$$F(z_1, z_2) = \begin{cases} (h_1 - z_1) + \dfrac{(\Delta h)^2 - (h_2 - z_2)^2}{2\Delta h} & h_1 > z_2, \ h_2 > h_1 > z_1 \\ \dfrac{(h_2 - z_1)^2 - (h_2 - z_2)^2}{2\Delta h} & h_1 < z_2, \ h_2 > h_1 > z_1 \end{cases} \tag{6.12}$$

where, as in [14], h_1 and h_2 are the minimum and maximum heights of built-up terrain in meter.

As in the first scenario, Eqs. (6.10)–(6.12) can be used for link budget design for various scenarios of the built-up terrain and for both BS and MS antennas lower than the rooftops.

6.2.1.2 Indoor propagation model
We use the following model of the indoor path loss developed in [13, 14] and adopt it for femto- and picocell scenarios, such as

$$L(r) = -L(r_0) - 10 \, n \log_{10} \left(\frac{r}{r_0} \right) - \sum \text{PAF} - \text{FAF} \tag{6.13}$$

where n is an exponent value coefficient, PAF (in dB) is a partition attenuation factor caused by a specific obstacle between the femto-access point and the mobile subscriber (MS), and FAF (in dB) is a floor attenuation factor. The values were selected according to the site's specific environments during the numerical simulation and according to the frequency band and appropriate empirical results specified in [13, 14].

We present the total path loss, given in [13, 14], accounting additionally for penetration of the radio signal inside buildings and the additional loss caused by walls, that is,

$$L_{\text{total}}(r) = -32.4 - 30 \log_{10} f_{(\text{MHz})} - 30 \log_{10} r_{(\text{km})} - L_{\text{fading}} - L_{\text{walls}}$$

$$+ (G_{\text{MS}} + G_{\text{BS}}) \tag{6.14}$$

accounting for fading phenomena, either formula (6.9) or (6.11) for the first or the second scenario, respectively; all other parameters are described above in this paragraph. Here, we also introduced a new term called loss by walls, $L_{\text{walls}} = 10 \log_{10} |T|$, $|T| = \sqrt{X} |\Gamma|)^2$, according to [25], where $|T|$ is the absolute value of refraction coefficient that equals $|T| = 0.9 - 0.95$ for glass, $|T| = 0.75 - 0.85$

for wood, $|T| = 0.55 - 0.65$ for stones, and $|T| = 0.15 - 0.2$ for concrete (for $X = 0.5 - 0.9$) [25, 27]. The wavelength of the radio wave has a wide range that varies from $\lambda = 0.05$ to 0.53 m and covers most of the modern wireless networks [14, 25–27].

We now consider an indoor scenario where the channels consist of crossing corridors and rooms that lie along each corridor. For such scenario we use a combination of the statistical waveguide model, describing propagation phenomena along the corridors and inside rooms lining each corridor (see [14]), in which instead of the real measured time t, we account for relations between the time $t = (r + \tilde{r})/d$ and distances r and $\tilde{r} = \sqrt{d^2 + r^2 - 2rd \cos \varphi}$. After straightforward derivations following [14], we get

$$L_{\text{total}} = -32.4 - 20 \log_{10} f_{\text{(MHz)}} - 10 \log_{10} \left[\frac{\gamma_0(r + \tilde{r})}{d} \sin^2 \frac{\alpha}{2} \right] - 2.4\gamma_0(r + \tilde{r}) -$$

$$10 \log_{10}(|T_{\text{wall}}| \cdot |T_{\text{floor}}|) - 10 \log_{10} \frac{d(r + \tilde{r} - d \cos \varphi)}{(r + \tilde{r})^2 - d^2} -$$

$$8.6 \frac{|\ln X|}{a'(\varphi)} \frac{(r + \tilde{r})^2 - d^2}{r + \tilde{r} - d \cos \varphi} + (G_{\text{MS}} + G_{\text{BS}}) \text{ dB} \tag{6.15}$$

In expression (6.15), we account for attenuation loss caused by walls and by floors, as was done in [25, 27], by introducing the absolute values of the coefficients of attenuation caused by the walls and floor according to [25]: $|T_{\text{wall}}| = \prod_m |T_{\text{wall},m}|$ and $|T_{\text{floor}}| = \prod_m |T_{\text{floor},m}|$. Formulas (6.8)–(6.12) will be used below for analysis of situation with path loss in the outdoor micro- and macrocell environments, and formulas (6.14) and (6.15) will be used for analysis of propagation inside femto- and picocell indoor links.

Equation (6.15) can be rewritten [14] to describe the path loss in the joint time-of-arrival (TOA) and angle-of-arrival (AOA) domains and differentiate the free space propagation loss from the guiding wave propagation loss along the corridor, that is:

$$L_t(t, \varphi) = 10 \log_{10}(I_{\text{total}}(\tau, \varphi)) = L_0 + L_{\text{su}}(t, \varphi) \tag{6.16}$$

where L_0 is the free space path loss, representing the attenuation along the direct path between the transmitter and the receiver, and L_{su} is a model of the path loss induced by the wave-guiding effect of the cylindrical waves propagation, or

$$L_{\text{su}}(t, \varphi) = 20 \log_{10} \sum_{n=1}^{n_{\text{max}}} \left\{ \left(\frac{1 - \chi |R_n(t, \varphi)|}{1 + \chi |R_n(t, \varphi)|} \right)^2 - \right.$$

$$\left. 8.6 \left[|\ln(\chi |R_n(t, \varphi)|)| \frac{d(\pi n - \varphi_n)}{a^2 + \rho_n^{(0)}} \right] \right\} \tag{6.17}$$

The first term in (6.17) represents the joint influence of the reflection coefficient and the waveguide discontinuity parameter γ on the waveguide-induced path loss. Note that in scenarios where the wave number n increases, the first term in (6.17) approaches zero. The second term in (6.17) represents the influence of the waveguide geometry on the path loss. In scenarios with large wave number n, the second term in (6.8) increases, and therefore the probability of the waveguide mode accumulation in the waveguide is decreased. Note that the path loss function along the corridor is obtained in [26] by taking into account the spatial and temporal signal interactions with the environment in the joint AOA–TOA domain. Figure 6.3, rearranged from [14], explains the phenomenon described above by presenting the wave-mode number n of the discrete spectrum.

It demonstrates the geometrical relation between the relative TOAs and AOAs as a function of propagating modes. The expressions for the relative TOA, τ, and AOA, φ, for the first three propagation modes are [14]

$$n = 1 \; \tau_1 = \frac{r_1 + \bar{r}_1}{d} = \frac{2\sqrt{(d/2)^2 + (a/2)^2}}{d} \quad \varphi_1 = \tan^{-1}\left(\frac{a}{d}\right) \quad (6.18)$$

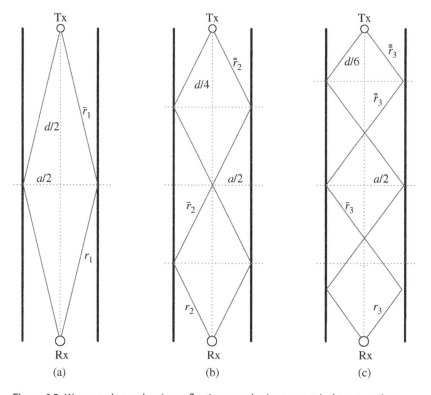

Figure 6.3 Wave-mode number (ray reflections number) n geometrical presentation: (a) $n = 1$, (b) $n = 2$, and (c) $n = 3$

$$n = 2 \ \tau_2 = \frac{r_2 + \bar{r}_2 + \bar{\bar{r}}_2}{d} = \frac{4\sqrt{(d/4)^2 + (a/2)^2}}{d} \qquad \varphi_2 = \tan^{-1}\left(\frac{2a}{d}\right)$$

$$(6.19)$$

$$n = 3 \ \tau_3 = \frac{r_3 + \bar{r}_3 + \bar{\bar{r}}_3 + \bar{\bar{\bar{r}}}_3}{d} = \frac{6\sqrt{(d/6)^2 + (a/2)^2}}{d} \qquad \varphi_3 = \tan^{-1}\left(\frac{3a}{d}\right)$$

$$(6.20)$$

Following similar arguments, the relative TOA and AOA of the nth wave mode (n reflections of the propagating rays in the corridor) can be derived as

$$n : \tau_n = \frac{2n\sqrt{\left(\frac{d}{2n}\right)^2 + \left(\frac{a}{2}\right)^2}}{d} \varphi_n = \tan^{-1}\left(\frac{an}{d}\right)$$

$$(6.21)$$

Let us present some examples of simulation of the total path loss L in decibels (dB) in the space domain (i.e. along the radio path between the transmitter and the receiver) according to (6.17) vs. the distance between the transmitter and the receiver. For our numerical computation we considered the following parameters: the width of the corridor $d = 3$ m, the conductivity of walls $\sigma = 0.0133$ S/m, and the signal frequency $f = 900$ MHz (according to the experiment described in [14]). The results of these path loss computations,

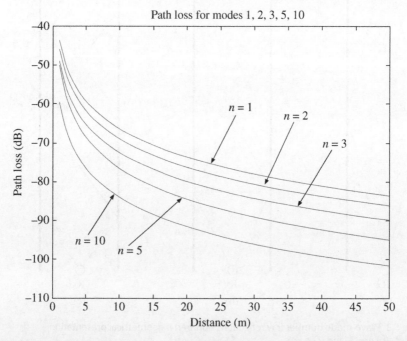

Figure 6.4 Path loss for $n = 1, 2, \ldots, 10$ wave modes vs. distance from the transmitter

according to (6.16), are shown in Figure 6.4 for the guiding modes, with the number n varying from 1 to 10. For $n \geq 3$, the effect of these modes is negligible at ranges beyond 20 m, and we just have to subtract the attenuation from the first two main modes of the original signal power in order to get the total power of a signal (in decibels) for each distance d between the transmitter and the receiver located along the corridor waveguide.

This effect was also shown in[20], where it was experimentally obtained that only one to two main modes are important in the range of 10 m or more from the transmitter. We will compare this theoretical prediction of the path loss with the real experiment in the following paragraph.

6.2.2 Experimental Verification of the Total Path Loss in Femtocell–Picocell Areas

Now we compare the results of numerical computations of propagation outdoor–indoor model for femtocell employments in the pico- and microcell environments with experiments carried out in a special built-up environment [14].

In the first cycle of experiments, we show only the part of the specific area topographic map that corresponds to a femtocell–picocell environment surrounding one of the two-story building, as shown in Figure 6.5, where the position of the transmitting antenna is denoted by the circle.

The vertically polarized transmitting sectoral antenna was installed 12 m apart from the building, at 2 m height (see Figure 6.5). The receiving antenna was positioned inside the building at a height of 5 m (i.e. at the second floor). It was arranged at the notebook as a wireless card with its dipole microantenna. This experiment corresponds to "femtocell–picocell" conditions.

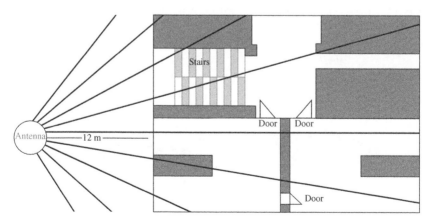

Figure 6.5 The scheme of the experimental site. The transmitter antenna is denoted by circle and located outside the building at the range of 12 m from the front wall of the building. Each line presents the radio path in azimuth domain where the angle is changed from 30° to 150°

Measurements were carried out for each meter along the radiopath outside and inside the building, and the corresponding signal strength was measured. The power of the transmitting signal was 12 dBm, and the frequency was 2.45 GHz. Due to scanning of the transmitting antenna (see the straight lines in Figure 6.5) in the azimuth domain, different angles of beam direction were taken, starting from 30° up to 150°, as seen from Figure 6.5. We present two characteristic graphs, obtained experimentally according to our 3D numerical code. The measured results are shown in Figure 6.6.

The corresponding numerical simulations of the same conditions of the experiment are shown in Figure 6.7a,b.

It was found, both theoretically and experimentally, that after passing a wall of bricks the signal intensity falls to 14–17 dB, and then attenuates smoothly according to expression (6.15) along the corridor. Despite the fact that the measured and the numerically predicted results of the signal 3-D pattern do not

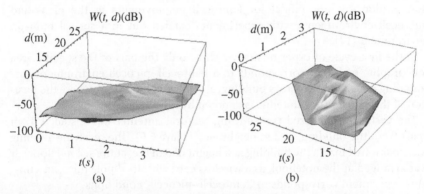

Figure 6.6 Experimentally obtained 3D pattern of the signal power in the joint time–distance domain for the azimuth of (a) 60° and (b) 135°

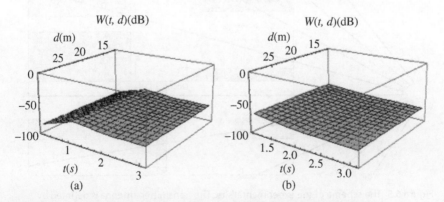

Figure 6.7 Numerical simulation outdoor–indoor link where all simulation data corresponds to the experiment presented in (a) Figure 6.6a and (b) Figure 6.6b

have the same shape and form, they both predict the sharp attenuation as the signal passes through the wall, and they predict the same attenuation (with accuracy of $\pm(3-5)$ dB).

Notice that inside the building, due to multidiffraction and multireflection effects from each inner architectural construction, the measured data shows strong oscillations of the recording signal strength (see Figure 6.6a,b), whereas the simulated data show much weaker oscillations (see Figure 6.7a,b).

Behind the building, the theoretical model is a poor predictor of the experimental data. The difference between theory and experiment is of the order of 10–15 dB. This occurs because in expression (6.16) the effects of attenuation due to several walls are not taken into account, as well as the effects of furniture and other architectural structures that can work as the "secondary sources of diffraction," increasing the overall intensity of the signal. Therefore, the multiparametric stochastic approach, presented above, is limited, as a good predictor of propagation phenomena in the indoor/outdoor femtocell–microcell communication environment, where all features and constructions, existing inside each room under testing, should be taken into account.

6.3 Different Integrated Femto/Pico/Micro/Macrocell Network Deployments

We analyzed numerically two types of deployments for different types of femtocell configurations: for four network configurations (CSG, OSG, SCA, and DCA) of the standard FAPs with maximum output power of 10, 15, and 21 dBm, described by (6.1)–(6.4).

6.3.1 Femtocells Integrated into Microcell Network Pattern

Numerical computations were performed for the urban scene that corresponds to one of the built-up areas of 1.5×1.5 km (see Figure 6.8), after the power allocation optimization algorithm using the water-filling mechanism. As it follows from Figure 6.8, there are seven cells uniformly distributed in this area. The MS users were randomly, but uniformly, distributed across the selected area where part of them was randomly allocated in the indoor environment according to the real clutter definition (that is, different percentages of indoor calls of 20%, 40%, and 60% were simulated according to conditions considered in [15, 16]).

The FAPs were also randomly uniformly distributed between the indoor users. Different percentages of indoor MS users (from all number of users located in area of service) of 50%, 70%, and 100%, which have FAPs, were also simulated. Different distributions of FAPs between the MS users were simulated for the following configurations: (i) uniform, where 80% MS users are concentrated at the cell edge and (ii) 20% are concentrated in the cell center, and vice versa.

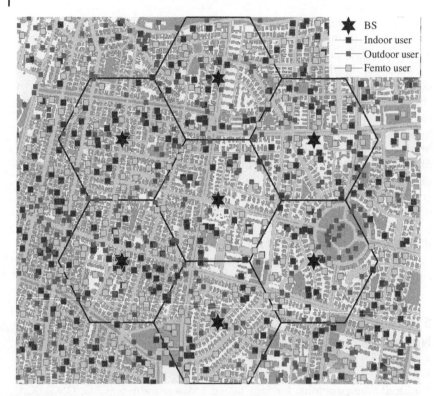

Figure 6.8 The tested urban area. The big stars correspond to the MBS; different collared quadrates correspond to femtousers, indoor users, and MS outdoor users, respectively

All types of FAP configurations (CSG and OSG) with SSA and DSA were simulated and are presented in Figures 6.9–6.12 for the abovementioned scenarios of MS users' distribution. The analysis was performed within the centrally positioned seven-cell pattern in the cluster assuming the frequency reuse of one. In numerical computations, the users' density was taken to be 480 users per square kilometer [15, 16]. Thus, in Figure 6.9, the CDF of signal data capacity for four network configurations (CSG, OSG, SCA, and DCA) and for different FAP distributions, 100% (a) and 50% (b), is presented.

The effective irradiated power is 21 dBm, and the number of indoor users is 60% from all subscribers in the tested area. One can see that the preferable network configurations are the shared and dedicated CSG and OSG femtonetwork configurations. Figure 6.10 shows the total probability of the signal capacity (or maximum data rate) for shared CSG (a) and dedicated OSG (b) for different FAP densities, of 50%, 70%, and 100%, for indoor users using FAP of 60% from the total number of indoor users in the area of service.

The dedicated OSG allows for achieving higher data rate for the same usage of femto-APs – 50%, 70%, or 100%. Figures 6.11 and 6.12 show the CDF of

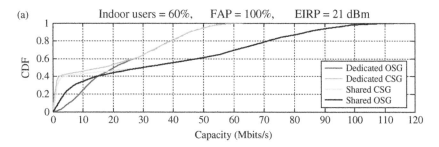

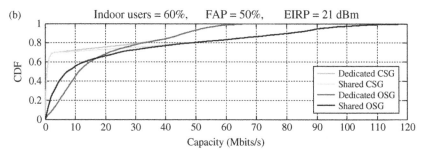

Figure 6.9 Distribution of signal data capacity for four network configurations (CSG, OSG, SCA, and DCA) for different FAP distributions: 100% (a) and 50% (b). The effective irradiated power is 21 dBm; the number of indoor users is 60% from all subscribers in the tested area

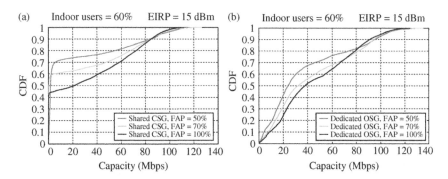

Figure 6.10 CDF of signal capacity for SOSG (a) and DOSG (b) for different FAP densities, of 50%, 70%, and 100%, for indoor users using FAP of 60% from the total number of users in the area of service

the signal capacity for the dedicated CSG and shared network configuration, respectively, for different percentages of indoor users, 20%, 40%, and 60%, and for number of FAPs used by these indoor users: 70% (a) and 100% (b).

We can see that, again, the shared CSG concept allows us to obtain higher data rate (or capacity) for the same number of FAPs deployed by indoor users.

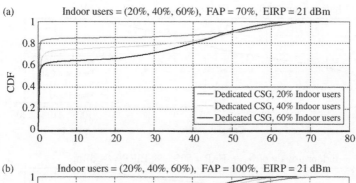

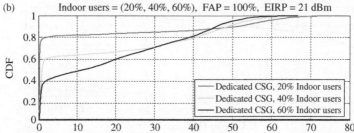

Figure 6.11 CDF of the signal capacity for the same dedicated CSG network configuration for different percentages of indoor users, 20%, 40%, and 60%, and for number of FAPs used by these indoor users: 70% (a) and 100% (b)

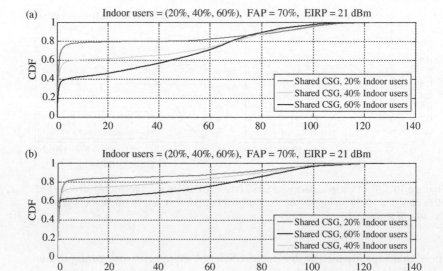

Figure 6.12 Network configurations vs. different percentages of indoor users, 20%, 40%, and 60%, for shared CSG scenario

The obtained results allow us to conclude that the water-filling techniques for the optimal power allocation of femto-APs can be fully implemented for predicting the different femtocell–macrocell network configurations with different trade-offs for indoor, outdoor, and femto-users densities. The similarly obtained results for two frameworks, with and without the water-filling mechanism, have shown an improvement in the capacity distribution results between the users of joint femtocell–macrocell coexisting systems.

6.3.2 Femto/Pico/Microcell Configuration Deployment

For small town (see Figure 6.13), we model different scenarios mentioned above with the different percentages of MS users arranged inside buildings, which have FAPs, that is, of 50%, 70%, 100%. Different positions of FAPs between MSusers were modeled: regular and homogeneous, with 80% of MSs located at the boundary of femtocell and 20% of MSs located at the center of femtocell, and conversely all deployments of FAPs – CSG and OSG for models from 3A to 3D – were analyzed (see definitions in the previous sections). All derivations during numerical experiment were carried out at the range of seven central cells (seven cells, see Figure 6.13) with the proposed reuse frequency algorithm. The users' layout was estimated roughly as ~ 460 users per (km^2).

Below, we analyze three regimes of femtocell deployment (A–C), described in Section 6.2: dedicated channel vs. cochannel; open access (OSG) vs. closed access (CSG); and fixed power in the downlink (DL) channel vs. adaptive DL power.

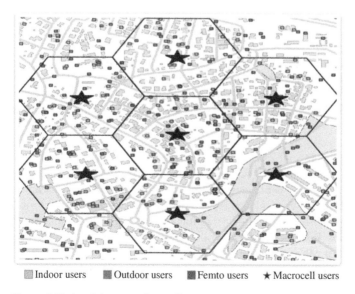

▦ Indoor users   ▩ Outdoor users   ▪ Femto users   ★ Macrocell users

Figure 6.13 A cellular map of a small town

Figure 6.13 presents the position of seven microcell stations at the map of the town with population of 50–60 thousands taken from [14], with femtocells "embedded" inside. Radio coverage of these seven microstations with layout of pico- and femtocell access points was evaluated according to the multiparametric stochastic model, depending on the corresponding channels' geometry described in Section 6.2 for outdoor and indoor scenarios, described by formulas (6.7)–(6.16), respectively. For this purpose, an RSSI of each received signal is derived for a map of indicator of each signal-received power performance, depending on its position with respect to BS (forward link) based on the topographic map of the area and the corresponding model of propagation, described by (6.8)–(6.12) and (6.13)–(6.15) for the corresponding outdoor and indoor scenarios, respectively.

Further steps of computations were related only with subscribers' layout, distributed inside desired pico- and femtocell areas of service according to the following algorithm:

- Capacity of each channel is derived for different configurations of the network described by (6.7)–(6.11) for all four configurations: CSG-SSA, CSG-DSA, OSG-SSA, and OSG-DSA.
- Each numerical experiment is carried out during a short period, that is, we assume that from the beginning we do not account for a difference between CSG and OSG, because it is related to the fact that from the beginning ($t = 0$), the amount of users for femto- and picocells is the same for both scenarios, CSG and OSG.
- For $t > 0$, for modeling of the difference between CSG and OSG, we should add a short intermediate stage accounting "handover stage" (switching), which is occurring at the further time period ($t + 1$) from the initial point. During this step we "give possibility" to each subscriber of OSG to enter in other femto- or picocells, according to their individual RSSI.
- All users are static, that is, during a short period of the desired scenario, any movements and changes of their position with respect to the desired femto- or picocells (RSSI does not change for each user during this short period).

We also checked a special scenario of networks based on picocells without incorporation of femtocells. This scenario is shown in Figure 6.14, where distribution of users' capacity is presented without femtocells, 50% of which are located inside buildings. Vertical axis represents the CDF (e.g. the total probability) of capacity per one user; the horizontal axis represents the capacity of each user in Mbps. Results clearly show that only for 13% of subscribers the capacity exceeds 0.4 Mbps (see Figure 6.14).

Then, we analyze the influence of deployment of FAPs inside the pico/micro network. For this purpose, two different approaches, described in Section 6.2, were introduced in consideration – FAPs in CSG and FAPs in OSG

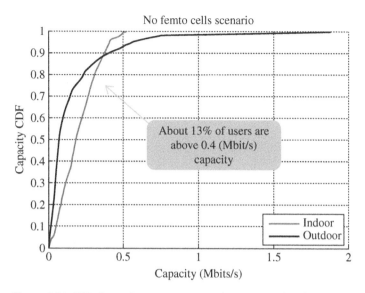

Figure 6.14 CDF of capacity per one user vs. the capacity of each user (in Mbps)

configuration. Figure 6.15a represents the first scenario (CSG), where in the first configuration, 10% of central subscribers and 80% of boundary femtocell (indoor) users were located. Figure 6.15b represents the second scenario (OSG), where in the second configuration, 10% of central subscribers and 80% of boundary femtocell (indoor) users were located.

Data presented in Figures 6.14 and 6.15 show that there is a sufficient increase in the capacity for each user using femtocells in pico/microcell configurations, and this tendency does not depend on the percentages of their deployment.

During the numerical computations it was found that each different scenario gave not so differing results in terms of user capacity. Therefore, estimating the obtained results, giving the optimal layout of femtocells for each separate configuration of the femtocell layout, was found to be a more effective approach. Therefore, the next step of analysis consists of more realistic scenarios of FAPs layout:

- the percentages of FAPs inside the buildings(25%, 50%, 75%),
- the percentages of FAPs in the center of the femtocells(25%, 50%, 75%, 100%),
- the percentages of FAPs at the edge of the femtocells(25%, 50%, 75%, 100%),
- the percentages of FAPs outside the femtocells(25%, 50%, 75%, 100%).

We had a total of 192 different variants (see Table 6.1), for each approach (CSG or OSG). Finally, the average results of computations for 10 different numerical experiments for each scenario were taken, that is, 1920 experiments for each condition of the network – dense and rare.

(a)

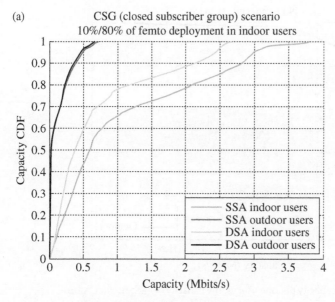

(b)

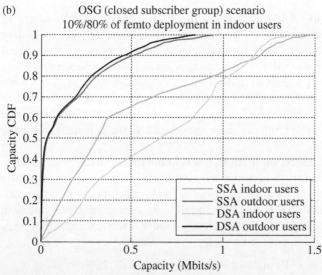

Figure 6.15 CDF of the capacity per one user vs. each user capacity (in Mbps) for four scenarios of users' allocation inside and outside buildings: (a) for CSG scenario and (b) for OSG scenario

Table 6.1 The parameters of modeled scenarios

Users inside buildings(%)	FAPs in center(%)	FAPs at edge(%)	FAPs outside femtocells(%)
25	25	25	25
25	25	25	50
25	25	25	75
25	25	25	100
25	25	50	25
. . .			
75	25	25	25
. . .			
75	100	100	100

Finally, 3840 numerical experiments (simulations) were carried out:

Users inside buildings (%)	25	25	25	25	25	75	75
Users inside buildings (%)	25	25	25	25	25	75	75
FAPs in the center of the femtocells (%)	25	25	25	25	25	25	100
FAPs at the edge of the femtocells (%)	25	25	25	25	50	25	100
FAPs outside the femtocells (%)	75	25	50	100	25	25	100

6.3.2.1 Results of the numerical computations

Based on the data of computation, the corresponding diagrams were performed (in numbers) for each desired network deployment and its configuration. We present below only those of them that have practical applications for the current regime of the femto/pico/microcell network (CSG-SSA, CSG-DSA, and so forth). Results include 64 points, 16 from each at the separate line, which determines a group of scenarios, having the same amount of FAPs in the center, and at the boundary of femtocells. Spreading of points along each line determines the limits of the capacity, depending on the searching scenario.

Thus, in Figure 6.16a,b the percentage of users in CSG-SSA scenario, 25% of which are located inside buildings, and the users' capacity distribution area is shown for FAPs located at the centers and at the edges of femtocells, respectively.

The scenario of 75% of users located inside buildings is shown in Figure 6.17a,b for the same configuration of FAPs – in the centers and at the edges, respectively.

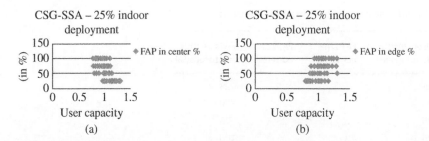

Figure 6.16 The percentage of users and their capacity distribution for the case of 25% of them located inside buildings: (a) FAPs in the center and (b) FAPs at the edges

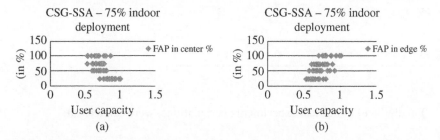

Figure 6.17 The same as in Figure 6.16a,b, but for the case of 75% of users located inside buildings

From Figures 6.16a and 6.17a it is clearly seen that an increase in the percentage of users inside buildings in femto/pico/micro cells leads to the decrease in the user's capacity (from 0.8–1.3 to 0.5–1.0 Mbps) for FAPs located in the center of femtocells. The same tendency of capacity decrease is observed for FAPs located at the edges of femtocells (see Figures 6.16b and 6.17b). It can be explained as follows: users located inside buildings have higher signal loss due to its passage through walls, which strongly decrease its power and, as result, the capacity. Therefore, for most users located inside buildings, lower mean values of the capacity are characterized. It can also be observed from the presented illustrations that more FAPs in the center lead to a decrease in the mean power obtained by each user, and more FAPs at the edges lead to an increase in the user's power. This occurs because the location of FAPs in the center of each microcell leads to an increase in interference noises and to a decrease in the signal power. Their deployment at the boundaries of cells (at the edges) will lead to a sufficient increase in average power for each subscriber located in the desired pico- or microcell. The same tendency is observed for open group (OSG-SSA) scenario.

As for the CSG-DSA scenario, the same tendency is observed for FAPs located at the centers of pico- or microcells (see Figure 6.18a). Difference is observed for users located at the boundaries and FAPS located at the edges of femtocells (see Figure 6.18b). For a more precise understanding of the

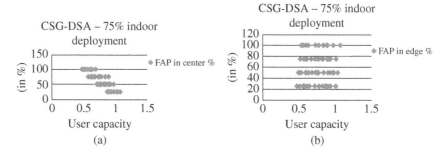

Figure 6.18 Percentage of users and their capacity (in Mbps) for CSG-DSA scenario when 75% of users are inside buildings and FAPs are (a) in the center and (b) at the edges

obtained results, it is necessary to differentiate two types of configurations: CSG/OSG-SSA and CSG/OSG-DSA. For the SSA approach (see Section 6.1), for all users the bandwidth is the same, that is, interference noises occur from all users' transmitters around the radius of the desired subscriber, whereas for the DSA approach, the user's bandwidth is a priori dedicated and the interference noises are significantly less.

For detail analysis of the corresponding formulas of the capacity for these two configurations, determined, for example, for indoor case, one should investigate the influence of the following parameters: I – interference, B_W – bandwidth, d – range between FAPs, and RSSI – indicator of the receiving power of the desired signal.

Thus, for SSA frequency allocation approach, for any user of femto- or picocell, the effects of interference, distance, and bandwidth remain the same for each user, that is, $I = $ const, $d = $ const, and $B_W = $ const, because users were allocated the same bandwidth. Here, only one parameter, RSSI, is changed, depending on the position of femto- or picouser with respect to BS antenna. We obtained in our simulation that the more femtousers are located at the boundaries of the corresponding cells, the higher total power of signal they can receive.

For DSA frequency allocation approach, the situation is more complicated. For the same distance ($d = $ const) for each femtouser, the interference I and bandwidth B_W have the tendency to change for different percentages of femtocell deployment. This occurs due to the dedicated bandwidth for each subscriber. With an increase in FAP amount deployment, the interference of picousers falls but that of femtousers increases, and conversely, with a decrease in FAP deployment, the interference of picousers increases but that of femtousers falls.

In addition, it was shown that an increase in the FAP amount decreases the capacity and B_W for femtousers and increases these characteristics for picousers. For DSA frequency allocation approach, a correlation was found between these three parameters, I, B_W, and RSSI. This yields insignificant

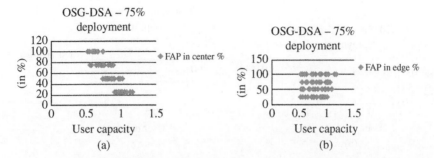

Figure 6.19 Percentage of users and their capacity per user (in Mbps) for 75% of users located inside buildings: (a) FAPs in center and (b) FAPs at the edges of femtocells

difference in diagrams presented above for the CSG scenario and can also be seen for the OSG scenario presented for DSA approach in Figure 6.19a,b, in which 75% of users are located inside buildings and FAPs are located in the center and at the edges of femtocells, respectively.

Results presented in Figures 6.18 and 6.19 show the same tendency for OSG-DSA and CSG-DSA configurations, as was mentioned above. The case where only 25% of users are located inside buildings is shown in Figure 6.20.

It is clearly seen from such a comparison that with a decrease in the amount of subscribers located inside buildings, there is an increase in the user's capacity (from 0.5–1.2 Mbps, shown in Figure 6.19a, to 0.9–1.5 Mbps, shown in Figure 6.20) for FAPs located in the center of femtocells.

The results of the numerical experiments, presented above for validation of the optimal configurations of modern networks, consisting of femtocells integrated into macrocells (Figure 6.8) and picocell/microcell (Figure 6.13), have shown that it is difficult to obtain any optimal configuration of femto/pico/micro/macrocell deployment. At the same time, during numerous variants analysis, important results were found, which can be summarized as follows:

- Amplification of general signal-to-interference-noise ratio (SNIR) inside the network does not always yield an increase in users' capacity, mostly in conditions of optimal deployment of FAPs inside pico- and microcells.

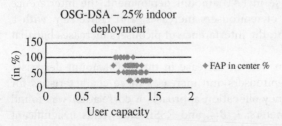

Figure 6.20 The same as in Figure 6.19a but for 25% of users located inside buildings

- A tendency in the network depending on the percentages of FAP deployment: lesser amount of users in the center of femtocells should use the corresponding FAPs. However, for users located at the boundaries of femtocells, the FAP deployment is optimal.
- Influence of the buildings' overlay profile on the capacity for each user and for the full capacity of the network [14, 16]: Higher buildings yield the worst scenario, called NLOS scenario. This effect is related to an increase in the spatial decay of radio signals and a decrease in the total capacity of each user's channel. Usage of femto antennas as the access points (FAPs) in such situations can increase the capacity of the designed network not only for each user located at boundaries of picocells but also for users located in their centers.

Finally, during numerical analysis it was shown that without detailed analysis of propagation conditions in the urban and/or suburban environments for different scenarios of BS and MS antenna locations with respect to buildings' overlay profile, occurring in the urban scene, it is impossible to design an effective structure in cellular map hierarchy and to predict the GoS and QoS for any subscriber located in the desired cell of cellular network – from femtocell to micro- and macrocell.

References

1 Chandrasekhar, V., Andrews, J.G., and Gatherer, A. (2003). Femtocell networks: a survey. *IEEE Commun. Mag.* 46(9):59–67.
2 Shannon, C.E. (1948). A mathematical theory of communication. *Bell System Tech. J.* 27:379–423 and 623–656.
3 Yeh, S.-P., Talwar, S., Lee, S.-C., and Kim, H. (2008). WiMAX femtocells: a perspective on network architecture, capacity, and coverage. *IEEE Commun. Mag.* 46(10):58–65.
4 Knisely, D.N., Yoshizawa, T., and Favichia, F. (2009). Standardization of femtocells in 3GPP. *IEEE Commun. Mag.* 47(9):68–75.
5 Knisely, D.N. and Favichia, F. (2009). Standardization of femtocells in 3GPP2. *IEEE Commun. Mag.* 47(9):76–82.
6 Chandrasekhar, V. and Andrews, J.G. (2009). Uplink capacity and interference avoidance for two-tier femtocell networks. *IEEE Trans. Wireless Commun.* 8(7):3498–3509.
7 Calin, D., Claussen, H., and Uzunalioglu, H. (2010). On femto deployment architectures and macrocell offloading benefits in joint macro-femto deployments. *IEEE Commun. Mag.* 48(1):26–32.
8 Kim, R.Y., Kwak, J.S., and Etemad, K. (2009). WiMAX femtocel: requirements, challenges, and solutions. *IEEE Commun. Mag.* 47(9):84–91.
9 Lopez-Perez, D., Valcarce, A., delaRoche, G., and Zhang, J. (2009). OFDMA femtocells: a roadmap on interference avoidance. *IEEE Commun. Mag.* 47(9):41–48.

10 Chandrasekhar, V., Andrews, J.G., Muharemovic, T. et al. (2009). Power control in two-tier femtocell networks. *IEEE Trans. Wireless Commun.* 8(8):4316–4328.

11 Yavuz, M., Meshkati, F., Nanda, S. et al. (2009). Interference management and performance analysis of UMTS/HSPA+femtocells. *IEEE Commun. Mag.* 47(9):102–109.

12 Femto Forum. http://www.femtoforum.org/femto/ (accessed 9 May 2020).

13 Blaunstein, N. and Christodoulou, C. (2007). *Radio Propagation and Adaptive Antennas for Wireless Communication Links*, 1e. Hoboken, New Jersey: Wiley.

14 Blaunstein, N. and Christodoulou, C. (2014). *Radio Propagation and Adaptive Antennas for Wireless Communication Networks–Terrestrial, Atmospheric and Ionospheric*, 2e. Hoboken, New Jersey: Wiley.

15 Blaunstein, N.Sh. and Sergeev, M.B. (2012). Definition of the channel capacity for femto-macrocell employment in urban environment with high layout of users. *J. Inf. Control Syst.* 3(58):54–62.

16 Tsalolihin, E., Bilik, I., Blaunstein, N., and Babich, Y. (2012). Channel capacity in mobile broadband heterogeneous networks based on femto cells. *Proceedings of EuCAP-2012 International Conference*, Prague, Czech Republic (26–30 March 2012), pp. 1–5.

17 Blaunstein, N. (1998). Average field attenuation in the non-regular impedance street waveguide. *IEEE Trans. Anten. Propag.* 46(12):1782–1789.

18 Blaunstein, N., Katz, D., Censor, D. et al. (2002). Prediction of loss characteristics in built-up areas with various buildings' overlay profiles. *J. Anten. Propag. Mag.* 44(1):181–192.

19 Yarkoni, N., Blaunstein, N., and Katz, D. (2007). Link budget and radio coverage design for various multipath urban communication links. *Radio Sci.* 42(2):412–427.

20 Ben Shimol, Y., Blaunstein, N., and Sergeev, M.B. (2015). Depolarization effects in various built-up environments. *Sci. J. Inf. Control Syst.* 69(2):83–94.

21 Blaunstein, N. and Levin, M. (1996). VHF/UHF wave attenuation in a city with regularly spaced buildings. *Radio Sci.* 31(2):313–323.

22 Blaunstein, N. (1999). Prediction of cellular characteristics for various urban environments. *J. Anten. Propag. Mag.* 41(6):135–145.

23 Okumura, Y., Ohmori, E., Kawano, T., and Fukuda, K. (1968). Field strength and its variability in the VHF and UHF land mobile radio service. *Rev. Electric. Commun. Lab.* 16(9–10):825–843.

24 Wells, P.J. (1977). The attenuation of UHF radio signal by houses. *IEEE Trans. Veh. Technol.* 26(4):358–362.

25 Bertoni, H.L. (2000). *Radio Propagation for Modern Wireless Systems*. Upper Saddle River, New Jersey: Prentice Hall PTR.

26 Seidel, S.Y. and Rappaport, T.S. (1992). 914 MHz path loss prediction models for indoor wireless communications in multifloored buildings. *IEEE Trans. Anten. Propag.* 40(2):200–217.

27 Yarkoni, N. and Blaunstein, N. (2006). Prediction of propagation characteristics in indoor radio communication environments. *J. Electromagn. Waves Appl.: Progr. Electromagn. Res., PIER* 59:151–174.

Part IV

Mega-Cell Satellite Networks–Current and Advanced

7

Advanced Multicarrier Diversity in Networks Beyond 4G

Diversity is a powerful communication receiver technique, which can be used to handle fading phenomena occurring in different wireless communication links, terrestrial, atmospheric, and ionospheric (described in [1–17]). Using diversity techniques, one can improve the performance of the multiple access system, operating in indoor/outdoor multipath environments. These techniques are based on the very simple principle of sending M copies of the desired signal data sequence (related to the desired user) via M different channels, instead of using only one channel to transmit and receive this desired information data.

There are different kinds of diversity techniques, which are currently used in canonical (e.g. current) and modern networks. We briefly described these techniques in Chapter 5, regarding the adaptive multibeam antenna applications. Here, we introduce some advanced techniques based on the proposed concept.

The analysis of fading, time- and frequency-varying, leads to the use of time-varying (adaptive) equalizers for stable communication achievement [3–7]. However, the design and use of time-varying and adaptive equalizers are difficult in practice, especially for broadband channels operating on the basis of adaptive/smart antennas. Only one solution currently exists, which is to use multicarrier (e.g. multichannel) techniques, based on frequency- and time-diversity algorithms or the space-diversity principle currently adapted for the multiple-input-multiple-output (MIMO) systems. This means that instead of one carrier, M carriers will be used to eliminate all kinds of noises, naturally or artificially generated.

7.1   Advanced Multicarrier-diversity Techniques

Before starting to analyze the methods of frequency, time, and space diversity, we determine quantitatively and show analytically the advantages that can

Advanced Technologies and Wireless Networks Beyond 4G, First Edition.
Nathan Blaunstein and Yehuda Ben-Shimol.
© 2021 John Wiley & Sons, Inc. Published 2021 by John Wiley & Sons, Inc.

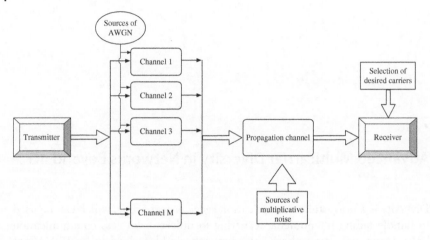

Figure 7.1 Multicarrier-diversity principle for desired user selection

be achieved using multicarrier-diversity methods. For this purpose, let us consider independent Rayleigh fading channels, i.e. operated in worst-case scenarios with the absence of line-of-sight (LOS) components, that is, for a fading parameter (see definitions in [3–11, 16, 17]). We call each channel a diversity branch. The corresponding scheme of how to combine and separate all carriers at the transmitter and then select the desired carrier at the receiver is shown in Figure 7.1.

We also assume that each branch has the same average signal-to-multiplicative noise ratio (SNR) defined as $\Gamma = E_b/N_0\langle\alpha^2\rangle$, where E_b is the energy of the bit of information data, passing the channel with strong fast fading, $\langle\alpha^2\rangle$ is the normalized deviation of signal data energy due to fading, and N_0 is the energy of the additive white Gaussian noise (AWGN). Each branch has an instantaneous SNR = γ_i. Then, the PDF of such a multiplicative noise, caused by fast and/or slow fading, can be written, for $\gamma_i \geq 0$, as

$$p(\gamma_i) = \frac{1}{\Gamma}\exp\left(-\frac{\gamma_i}{\Gamma}\right) \tag{7.1}$$

where Γ is the average SNR for each branch.

The probability that a single branch has SNR less than some threshold of multiplicative noise, γ, is the cumulative distribution function (CDF) defined for Rayleigh channel with fading as

$$\text{CDF}(\gamma_i) \equiv P(\Gamma_i \leq \gamma) = \int_0^\gamma p(\gamma_i)d\gamma_i =$$

$$\int_0^\gamma \frac{1}{\Gamma}\exp\left(-\frac{\gamma_i}{\Gamma}\right)d\gamma_i = 1 - \exp\left(-\frac{\gamma_i}{\Gamma}\right) \tag{7.2}$$

Now, the probability that all M independent diversity branches receive signals which are simultaneously less than some specific SNR threshold γ can be presented as

$$P(\gamma_1 \leq \gamma, \gamma_2 \leq \gamma, \dots, \gamma_M \leq \gamma) = \left[1 - \exp\left(-\frac{\gamma_i}{\Gamma}\right)\right]^M \equiv P_M(\gamma) \qquad (7.3)$$

This probability describes situation when all branches cannot achieve this threshold level γ. If any single branch with index i achieves this threshold, i.e. $\text{SNR}_i \geq \gamma$, then

$$P(\gamma_i > \gamma) = 1 - P_M(\gamma) = 1 - \left[1 - \exp\left(-\frac{\gamma_i}{\Gamma}\right)\right]^M \qquad (7.4)$$

Expression (7.4) describes the situation of exceeding of the threshold when selection diversity is used. For low Γ, that is, for strong fading effects, (7.4) reduces to

$$P \propto \left(\frac{\gamma}{\Gamma}\right)^M \qquad (7.5)$$

For selection diversity, the PDF is found as the derivative of CDF of all branches, to achieve threshold, that is,

$$p_M(\gamma) = \frac{dP_M(\gamma)}{d\gamma} = \frac{M}{\Gamma}\left[1 - \exp\left(-\frac{\gamma}{\Gamma}\right)\right]^{M-1} \exp(-\gamma/\Gamma) \qquad (7.6)$$

Then, the mean SNR, $\bar{\gamma}$, can be defined as

$$\bar{\gamma} = \int_0^\infty \gamma \cdot p_M(\gamma)d\gamma = \Gamma \int_0^\infty Mx(1 - e^{-x})^{M-1}dx \qquad (7.7)$$

where $x = \gamma/\Gamma$. Expression (7.7) is evaluated to obtain the average SNR improvement, offered by the selection diversity

$$\frac{\bar{\gamma}}{\Gamma} = \sum_{k=1}^{M} \frac{1}{k} \qquad (7.8)$$

As was shown by numerous computations, for the independent Rayleigh fading branches, as channels with the average multiplicative noise, the probability that the SNR drops below some specific threshold for one branch is two to three times greater in magnitude than if several independent (i.e. separated) branches are used in the multicarrier-diversity technique.

7.2 Advanced Frequency Multicarrier-diversity Techniques

This advanced technique allows modulating the information data signal (e.g. the baseband signal) through different M carriers. Frequency diversity allows

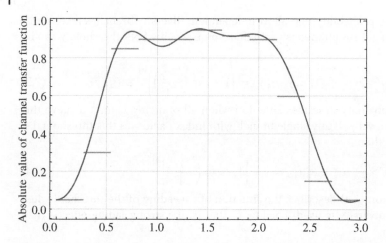

Figure 7.2 The principle of splitting of the whole channel bandwidth on *N* subchannels where effects of signal deviations are minimal

the transmission of information on more than one carrier frequency. Here, frequencies are separated by less (or equal) to the coherent bandwidth of the channel and, therefore, will not experience the frequency- or time-selective fades (see definitions above). We need to choose the symbol duration T_s such that the coherence bandwidth of each subchannel (denoted as $b_w \equiv b_c = B_c/N$; N is the number of carriers) will be much smaller than the bandwidth of the channel, B_W. In this case, slow and flat fading will take place, and the effects of frequency-selective fast fading will be minimized. The above assumption leads to the following constraint:

$$B_W > \frac{1}{T_s} \equiv b_c \tag{7.9}$$

Using N carriers, we finally have

$$N = \frac{B_W}{b_c} = \frac{B_W}{1/T_S} = B_W \cdot T_s \tag{7.10}$$

All the definitions of the parameters are presented in Section 7.1.

Figure 7.2 depicts an example of how we can split the channel bandwidth B_W into N subchannels with a bandwidth b_c which is narrow enough to exclude the effects of deep fading and narrowband intercarrier interference (ICI). Then, each independent symbol signal will have, in frequency domain, a rectangular shape of power spectral density (PSD), which, in the time domain, has a sharp δ-function presentation (see Figure 7.3a). Conversely, the rectangular shaping function $g(t)$ in the time domain (i.e. a pulse with data) has, in the frequency domain, a shape of the sinc function (see Figure 7.3b) that in the literature is called the Nyquist-shaped filter or the ideal filter [1–7].

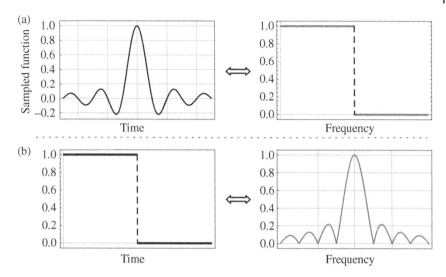

Figure 7.3 (a) The sinc function in the time domain (*t*) of the subsignal (left panel) and its rectangular shape in the frequency domain (*f*) (right panel). (b)The sinc function in the frequency domain of the subsignal (right panel) and its rectangular shape in the time domain (left panel)

7.3 Advanced OFDM and OFDMA Technologies

As follows from the description of frequency-division duplexing (FDD) and frequency-division multiple access (FDMA) techniques, described in [16], the spectral efficiency of the above techniques is too weak, because of the existence of guard intervals (i.e. the loss of useful bandwidth spectra) [2–10]. To eliminate this problem, it is more effective to use the independent (e.g. orthogonal) subcarriers, as shown in Figure 7.4.

Figure 7.4 The spectral overlapping for each subcarrier in OFDM technique, when each peak of any subcarrier corresponds to zeros positions of the other subcarriers due to their orthogonality

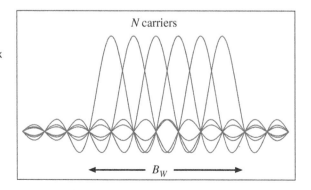

In such technique, one can split the total frequency-band spectra of the system to separate overlapping subcarriers (subbands or subchannels) with independent properties. This technique is called the *orthogonal frequency-division multiplexing* (OFDM) for resolving the problem of spectral overlapping, because each individual subcarrier, being orthogonal with respect to other subcarriers, can be easily recovered despite the overlapping in the total spectra. Thus, there is no need for guard intervals as used in FDD or FDMA techniques [2–10, 16].

7.3.1   Orthogonal Frequency-Division Multiplexing

Let us now consider the OFDM procedure as a pure mathematical problem. We assume that the transmitted signal passing the fading channel consists of N subcarriers (corresponding to N paths) and can be generally presented using the fading factor introduced above. Therefore, a total data signal (e.g. baseband signal) can be presented as a function of the amplitude of the signal received in the nth subchannel, denoted by α_n, and its own phase, $\Delta\varphi_n = n\,\Delta\omega\,t$, as

$$S(t) = \sum_{n=0}^{N-1} \alpha_n \cos[(\omega_0 + n\Delta\omega)t + \varphi_n] \tag{7.11}$$

where

$$\omega_0 = \frac{2\pi L}{T_s} \qquad \Delta\omega = \frac{2\pi}{T_s} \tag{7.12}$$

and L is the length of the channels.

The subcarriers of OFDM are orthogonal on the interval $[0, T_s]$, from which it follows that

$$\int_0^{T_s} s(t)\cos[(\omega_0 + n\Delta\omega)t + \varphi_n]dt = \alpha_n T_s \cos\varphi_n$$
$$\int_0^{T_s} s(t)\sin[(\omega_0 + n\Delta\omega)t + \varphi_n]dt = \alpha_n T_s \sin\varphi_n \tag{7.13}$$

The corresponding splitting allows us to obtain a signal for each carrier in the following manner:

$$s^{(q)}(t) = A_c \cdot \sum_{n=0}^{N-1} \alpha_n^{(q)} \cos[((\omega_0 + n\Delta\omega)t + \varphi_n^{(q)})], \quad (q-1)T_s < t < q\,T_s \tag{7.14}$$

where

$$a_n^{(q)} \equiv \alpha_n^{(q)} \exp(-j\varphi_n) \tag{7.15}$$

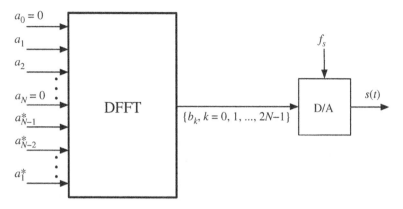

Figure 7.5 Block scheme of the fast Fourier transform technique, inverse (IFFT) at the transmitter and direct (DFFT) at the receiver; D/A is the digital-to-analog converter

$a_n^{(q)} \in A$, and A is a set of constellation points containing L points

$$\mathbf{a}^{(q)} \equiv (a_0^{(q)}, a_1^{(q)}, \dots, a_{N-1}^{(q)}) \tag{7.16}$$

or for orthogonal (independent) subchannels $a_n^{(q)} \in A, n = 0, 1, \dots, N - 1$.

Since its beginning, the OFDM-technique implementation was based on the discrete Fourier transform (DFT), mathematically described by (7.11) or (7.14) [11–15]. Simply speaking, DFT converts the time-domain representation of a signal with data to a frequency-domain representation. Conversely, the inverse discrete Fourier transform (IDFT) converts the signal spectrum, that is, the frequency-domain signal data representation to the time-domain representation. Later, instead of the DFT/IDFT technique, the direct and inverse fast Fourier transforms (denoted DFFT and IFFT, respectively) were used to significantly decrease the implementation complexity and time of the proposed technique. Mathematically, both methods are similar, but the FFT implementation is much more efficient. Below we briefly present the mathematical aspects of the FFT technique for the OFDM implementation.

First of all, we should state that the corresponding block diagrams of the IFFT for the transmitter and of the DFFT for the receiver are circuitwise and include similar blocks (only the block of digital-to-analog (D/A) should be changed to analog-to-digital (A/D), correspondingly). We present in Figure 7.5 the block-diagram of the receiver, where the samples of the multicarrier signal can be obtained by the DFFT of the data symbols.

According to the key goal of the OFDM modulation technique, the corresponding discrete-form presentation of the IFFT algorithm at the transmitter is the following: a sequence of the discrete signals with the noise, $\{b_k\}$, for each independent subcarrier (or subchannel), is presented in the following manner (following Figure 7.5, $a_0 = a_n = 0$):

$$b_k = \frac{1}{\sqrt{2N}} \sum_{n=0}^{2N-1} a_n \exp\left(j\frac{2\pi nk}{2N}\right)$$

$$= \frac{1}{\sqrt{2N}} \left[\sum_{n=0}^{N-1} a_n \exp\left(j\frac{2\pi nk}{2N}\right) + \sum_{n=N}^{2N-1} a_{2N-n}^* \exp\left(j\frac{2\pi nk}{2N}\right)\right]$$

$$= \frac{1}{\sqrt{2N}} \left[\sum_{n=1}^{N-1} a_n \exp\left(j\frac{2\pi nk}{2N}\right) + \sum_{m=1}^{N-1} a_m^* \exp\left(j\frac{2\pi(2N-m)k}{2N}\right)\right]$$

$$(7.17)$$

where for the second sum, we substituted $m = 2N - n$. In (7.17), the amplitude of each subcarrier can be presented in the baseband form:

$$a_n = \alpha_n \exp(j\phi_n) \tag{7.18}$$

Finally, using direct FFT (DFFT), we get

$$b_k = \frac{1}{\sqrt{2N}} \sum_{n=0}^{N-1} \alpha_n \left\{\exp\left[j\left(\frac{2\pi nk}{2N} + \varphi_n\right)\right] + \exp\left[-j\left(\frac{2\pi nk}{2N} + \varphi_n\right)\right]\right\}$$

$$= \frac{2}{\sqrt{2N}} \sum_{n=0}^{N-1} \alpha_n \cos\left(\frac{2\pi nk}{2N} + \varphi_n\right) \tag{7.19}$$

One may ask: "what does the introduction of the second term in expression (7.17) mean? (that is, the extension of the OFDM symbol sequence by introducing the second term)." If apriori, due to the multiplicative noise occurring in the multipath channel with fading (defined by the maximum delay spread), the previous part of each OFDM symbol will be corrupted by the delay sample of the neighboring OFDM symbol, the orthogonality between symbols will be lost, leading to the so-called intersymbol interference (ISI) or ICI [10–14, 17].

Since OFDM technique excludes the usage of the guard intervals compared to the FDMA system (see Ref. [17]), it is possible to extend the OFDM symbol sequence with additional replica consisting of N symbols of "zeros" corresponding to the so-called virtual guard with duration of T_g.

Using this IFFT technique, described mathematically by expression (7.17), we obtain that the first term in (7.17) will present a desired signal of symbol data, from which at the receiver, the transmitted symbol sequence $\{b_k\}$ can be easily recorded. The second term in expression (7.17) corresponds to the part of the symbols that can be corrupted by fading during passing via the communication subchannel. In other words, the second sequence of samples will be transmitted in the guard period (see Figure 7.6) as a cycle process. Therefore, this sequence in the literature is called cycle prefix [2, 7]. This extracted sequence will finally be eliminated at the receiver, if the elements of the second sum in formula (7.17) will be substituted by "zeros" during T_g.

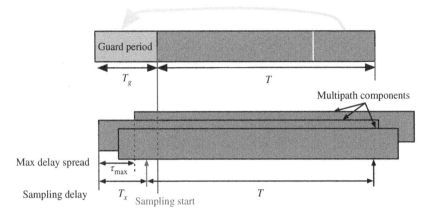

Figure 7.6 The procedure of obeying the part of the data of OFDM signal corrupted by fading (defined by the sampling delay time) by extension of the symbol time *T* with the "virtual" guard period (cyclic prefix)

If so, we can convert the symbol sequence in the time domain at the transmitter using IFFT in such a manner that its time period T_s will be extended on T_g, that is, $T = T_s + (T_g - T_x)$ (see Figure 7.6). This procedure is called the prefix cycling [2, 7]. After such a procedure, we can obtain expression (7.19) as a real symbol replica recorded at the receiver after implementation of the DFFT procedure. Using the above notations, we finally present for each independent subchannel the signal-shaped function at the receiver in the following form, using just a DFFT on the sampled symbol signal $s(t)$:

$$
\begin{aligned}
s(k\Delta t) &= \sum_{n=1}^{N-1} \alpha_n \cos(2\pi n \Delta f k \Delta t \varphi_n) \\
&= \sum_{n=1}^{N-1} \alpha_n \cos\left(2\pi n \Delta f k \frac{1}{2N\Delta f} + \varphi_n\right) = \frac{\sqrt{2N}}{2} b_k
\end{aligned}
\tag{7.20}
$$

In (7.20), the following parameters are introduced:

$$
\Delta f = \frac{1}{T_s}, \quad \Delta T = \frac{Ts}{2N} = \frac{1}{2N\Delta f}, \quad f_s = \frac{1}{\Delta t}
$$

Indeed, at the receiver, the N independent copies are combined in such a manner to give an optimal replica of the signals with data sequences of samples that are not corrupted by fading. Unfortunately, what is easy to perform using mathematical algorithms cannot be ideally obtained in practice of wireless communication, where the subchannel time-scale (or length) is not constant, and therefore the preface cycle parameters are not constant as well.

Another problem that should be avoided using the OFDM technique is the frequency shift of the received signal spectrum called the frequency offset [2, 7]. Due to this effect, the IFFT procedure at the transmitter and the DFFT

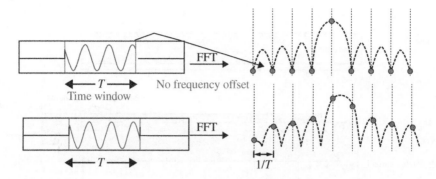

Figure 7.7 Effects of frequency offset on OFDM FFT modulation technique

procedure at the receiver are not "symmetrical," i.e. they do not correspond strictly to each other. The effect of the frequency offset is clearly seen in Figure 7.7.

According to frequency shifting, adjacent subcarriers can be affected by ICI, caused by the energy leakage of the neighbor symbol signals to each other. As was shown in [2, 7], the overall energy of ICI grows with the frequency offset. To avoid this effect, a frequency-domain equalization using separate equalizers for each subcarrier is needed and significantly increases the complexity of the receiver implementation. For further reading on this approach, we refer the reader to Refs. [7–9, 12–15].

To complete our explanation of the OFDM technique, we should mention that before starting the splitting of the specific channel into subchannels, the coherent bandwidth of the subchannel (we assume that they have the same bandwidth $\Delta f_n \equiv f_0$, $n \in \{1, 2, \dots, N-1\}$) should be estimated apriori. As described in [16, 17], during a special campaign carried out in the city of Tokyo, two scenarios were tested: one of heavy buildings layout (the first scenario) and the other of a lower building layout (the second scenario). Following the measurements, the first scenario of a range of 70–90 kHz with an average value of $b_{cn} \approx 80$ kHz was obtained, whereas for the second scenario, $b_{cn} \approx 390$–400 kHz. If now, after the OFDM-division procedure on N subchannels, the bandwidth f_0 of each subcarrier would be set to $f_0 = 100$ kHz, as follows from the second scenario, f_0 is smaller than the corresponding bandwidth b_{cn}, and the OFDM procedure fully obeys the fading phenomena in each subchannel for the second scenario. However, such a division procedure is not effective in obeying any fading phenomena for the first scenario, where $f_0 \geq b_{cn}$.

Finally, a strong frequency-selective fading occurring in a dense urban scene can affect each subchannel by corrupting the signal data for each subscriber located in the area of service. This example emphasizes the fact that before using the OFDM procedure, each designer of wireless network should estimate

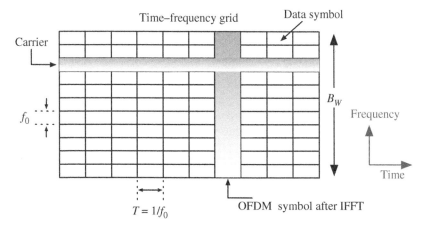

Figure 7.8 2D time–frequency signal presentation according to OFDMA technique

the fading parameters for each urban scenario, such as the time delay spread and coherence bandwidth (for stationary channels) and the Doppler shift bandwidth and time of coherency (for dynamic channels).

7.3.2 Orthogonal Frequency-Division Multiple Access

From Eq. (7.20), it is clearly seen that the OFDM is a one-dimensional (1-D) technique because the branches are splitting only in the frequency domain. A new multiple access technique was introduced during the first decade of the second century (see Refs. [4–8, 13–15]), which is two-dimensional (2-D) splitting of the signal in bins both in time and frequency domain. This multicarrier access procedure is called orthogonal frequency-division multiple access (OFDMA). Here, we briefly explain its algorithm and difference with respect to standard OFDM modulation technique. For this purpose, we arrange the corresponding scheme of each carrier signal-data presentation in the joint 2D time–frequency domain, as shown in Figure 7.8. One may notice that in an OFDMA system, for each carrier we get the narrow bandwidth $f_0 = B_W/N$ with the intercarrier separation that equals $1/T_s \ll f_0$.

We should also mention that during this procedure of splitting the total bandwidth B_W into N subchannels, each carrier bandwidth f_0 must be smaller than the bandwidth of coherency of each subchannel, that is, $f_0 < b_{cn}$, as was proved experimentally in [16, 17]. Considering the above constrain, we can, using the OFDMA technique, fully exclude the ICI, that is, the overlapping between each separate bin and, finally, not to spend resources (i.e. bits) for guarding effects, as was used in narrowband technologies, such as the FDMA and TDMA [2]. Therefore, the OFDMA technique can be considered as a hybrid FDMA/TDMA scheme described in [17], because it allows users to flexibly share both the frequency subband (e.g. the carrier) and the time slots. Following the above, we state the following features of OFDMA:

- It is based on OFDM in the sense that the multiple narrow-band subcarriers are modulated in parallel;
- It combines OFDM modulation and a multiple access scheme, say TDMA, as is shown in Figure 7.9;
- It combines time- and frequency-division multiple access techniques, that is, OFDMA=TDMA+FDMA (see Figure 7.10).

Moreover, the above analysis allows us to notice that using a large number of parallel narrowband subcarriers instead of a single wideband carrier to transport information, we can

- easily and efficiently deal with time-dispersive multipath fading;
- protect against narrow-band interference due to the orthogonality of the subcarrier channels;
- offer the flexibility to adapt the transmission rate per narrowband subchannel (e.g. subcarrier) to the most suitable transmission electronic schemes at the transmitter;

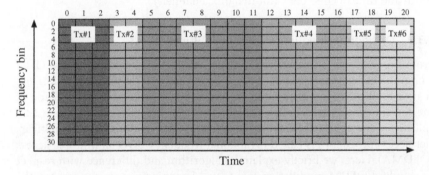

Figure 7.9 OFDMA as a combination of OFDM modulation and TDMA technology

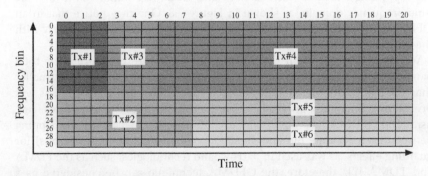

Figure 7.10 OFDMA as a combination OFDM modulation simultaneously with FDMA and TDMA technologies

- reduce some of the electronic elements at the receiver, because using such a technique, we do not need to implement the N oscillators, filters, and so forth, for each carrier.

At the same time, we notice that the OFDM/OFDMA techniques have some disadvantages of this technique, such as

- This technique is sensitive to frequency and phase offsets;
- It has a peak-to-average power ratio (PAPR) problem that reduces the power efficiency of the RF amplifier at the transmitter (for any multicarrier technique).

Despite the fact that the OFDM/OFDMA techniques have the above disadvantages, they were taken as strong candidates for most of the 3-G and 4-G wireless networks and were adapted in various modern standard technologies, such as the wireless personal area networking (WPAN or Bluetooth), the wireless local access networking (WLAN), which is equivalent to wireless fidelity (Wi-Fi), the worldwide interoperability for microwave access(WiMAX), the long-term evolution (LTE), and so on. Currently, the OFDMA technology is intensively used in IEEE 802.16/WiMAX standard networks and in their combination with MIMO systems (see the brief descriptions of the corresponding networks in Section 7.4).

7.4 Advanced Time Multicarrier-diversity Techniques

As was defined in [16, 17], in situations when the signal bandwidth is larger than the coherence bandwidth of the channel, that is, $B_W \gg B_c$, the channel is frequency selective and fast. If the same channel is subdivided into a number of orthogonal frequency-division multiplexing subchannels having a mutual separation in center frequencies of at least $\Delta f = B_c$, the effects of fading on the signal data transmitted via each subchannel can be eliminated.

However, in the case of wideband modulation, such as in the CDMA technique, the multiple sequences of the data signal from each subscriber arriving at the receiver can destroy the independence between the codes (i.e. their orthogonal properties) if their delays will exceed a single chip duration. This usually occurs if the chip rate exceeds the coherent bandwidth of the subchannel, that is, $R_c > b_c$. In such scenarios, an alternative method is usually used on the basis of a so-called RAKE detector [3–5]. Here, a time-diversity technique is used, assuming $T_s \gg T_c = 1/b_c$, that is, the information signals duration exceeds the coherence time of the subchannel. In this case, multiple repetitions of the signal will be received with independent fading conditions. In other words, we can obtain time diversity by transmitting the same signal multiple times and separated signals apart in time in such a manner that the channel multipath fading will be uncorrelated between replicas.

As was shown in references [3–5], the one modern implementation of time diversity involves the use of a RAKE receiver, working as n-delay line through which the received signal is passed. Its action is somewhat analogous to an ordinary garden rake, and consequently, the name "RAKE receiver" has been used for this device by Price and Green in 1958 [3–5].

As mentioned in Chapter 1 (see also Ref. [17]), in DS-SS system (or CDMA), the chip rate is much greater than the fading bandwidth of the channel. CDMA spreading PN codes are designed to provide very low correlation between successive chips. Thus, propagation delay spread in the wireless channel merely provides multiple versions of the transmitted signal at the receiver. If these multipath components are delayed in time by more than the chip duration, they appear like uncorrelated noise at the CDMA receiver, and an equalizer is not required. However, since there is useful information in the multipath components, CDMA receivers can combine the time-delayed versions of the original signal transmission in order to improve the SNR at the receiver. The RAKE receiver is usually used for such purposes. This receiver, using the tapped delay line structure with discrete time intervals equal to the chip period T_c, multiplies several copies of the received signal by versions of a spreading code, shifted by multiples of T_c. A RAKE detector collects the time-shifted versions of the original signal by providing a separate correlation receiver for each of the multipath signals. The RAKE receiver, shown in Figure 7.11, is essentially a diversity receiver designed especially for CDMA, where the diversity is provided by the fact that the multipath components are practically uncorrelated from one another, when their relative propagation delays exceed a chip period T_c.

As seen from Figure 7.11, a RAKE receiver utilizes multiple correlators to separately detect the M multipath components with deep fading (i.e. the strongest multipath components). Such a procedure allows the components of the desired signal with data (e.g. with an original bit sequence) to be recovered and recombined with the corresponding time shifts due to the channel removing.

Let us briefly analyze the RAKE receiver working process in more detail following the block scheme presented in Figure 7.11. The outputs of each correlator are weighted to provide a better estimate of the transmitted signal.

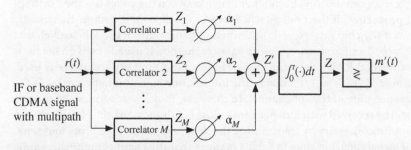

Figure 7.11 Block scheme of the RAKE receiver

Demodulation and bit decisions are then based on the weighted outputs of the M correlators. Assume that M correlators are used in a CDMA receiver to capture the M strongest multipath components. A weighting network is used to provide a linear combination of the correlator output for bit detection.

The first correlator is synchronized to the strongest multipath component m_1 of the signal $r(t)$ with data and multiplicative noise due to fading. The multipath component m_2 arrives τ_1 later than the component m_1. The second correlator is synchronized to the component m_2. It is correlated strongly with m_2 but has low correlation with m_1. Note that if only one correlator is used, as is usually done for single-carrier systems, such as FDMA and TDMA, the strong fading corrupting the channel cannot be eliminated by a single receiver. Then bit decisions based on only a single correlation may result with a large bit error rate of the information data passing through such a single-carrier channel. In a RAKE receiver, if the output from one correlator is corrupted by fading, the others may not, and the corrupted signal may be discounted through the weighting process. Decisions based on the combination of the M separate decision statistics offered by the RAKE provide a form of diversity which can overcome fading and thereby improve CDMA reception. The M decision statistics are weighted to form an overall-decision statistics, as shown in Figure 7.11.

The outputs of the M correlators are denoted as $Z_1, \ldots, Z_M$ and are weighted by the coefficients $\alpha_1, \ldots, \alpha_M$, respectively. The weighting coefficients are based on the power or the SNR from each correlator output. If the power (or SNR) is small at the output of a particular correlator, it will be assigned a small weighting factor. As in the combining diversity scheme, the overall signal Z' is given as

$$Z' = \sum_{m=1}^{M} \alpha_m Z_m \tag{7.21}$$

The weighting coefficients, α_m, are normalized to the output signal power of the correlator in such a way that the coefficients sum is equal to 1, shown by the following formula:

$$\alpha_m = \frac{Z_m^2}{\sum_{j=1}^{M} Z_j^2} \tag{7.22}$$

As in the case of adaptive equalizers and diversity combining, there are many ways to generate weighting coefficients. Choosing weighting coefficients based on the actual outputs of the correlators yields better RAKE receiver performance. This performance gives a conditional error probability in the form of

$$P(\gamma_b) = Q\left(\sqrt{\gamma_b(1 - \rho_r)}\right) \tag{7.23}$$

where $\rho_r = 0$ for orthogonal signals and $\rho_r = -1$ for antipodal signals; the Q-function was defined in the literature via the error function [3–17]. Here, γ_b

is the current SNR, which equals

$$\gamma_b = \frac{E_b}{N_0} \sum_{k=1}^{M} \alpha_k^2 = \sum_{k=1}^{M} \gamma_k \tag{7.24}$$

For Rayleigh fading channel, we can finally obtain the probability for instantiations SNR, γ_k:

$$P(\gamma_k) = \frac{1}{\overline{\gamma_k}} \exp\left(-\frac{\gamma_k}{\overline{\gamma_k}}\right) \tag{7.25}$$

where, as above, $\overline{\gamma_k} = E_b\langle\alpha_k^2\rangle/N_0$ is the average SNR for the kth path (kth subcarrier or subchannel).

References

1 Steele, R. (1992). *Mobile Radio Communication*. IEEE Press.
2 Proakis, J.G. (1995). *Digital Communications*, 3e. New York: McGraw-Hill.
3 Stuber, G.L. (1996). *Principles of Mobile Communications*. Boston, MA: Kluwer Academic Publishers.
4 Rappaport, T.S. (1996). *Wireless Communications: Principles and Practice*, 2e in 2001. New York: Prentice Hall PTR.
5 Steele, R. and Hanzo, L. (1999). *Mobile Communications*, 2e. Chichester: Wiley.
6 Li, J.S. and Miller, L.E. (1998). *CDMA Systems Engineering Handbook*. Boston, MA and London: Artech House.
7 Burr, A. (2001). *Modulation and Coding for Wireless Communications*. New York: Prentice Hall PTR.
8 Paetzold, M. (2002). *Mobile Fading Channels: Modeling, Analysis, and Simulation*. Chichester: Wiley.
9 Simon, M.K., Omura, J.K., Scholtz, R.A., and Levitt, B.K. (1994). *Spread Spectrum Communications Handbook*. New York: McGraw-Hill.
10 Glisic, S. and Vucetic, B. (1997). *Spread Spectrum CDMA Systems for Wireless Communications*. Boston, MA and London: Artech House.
11 Dixon, R.C. (1994). *Spread Spectrum Systems with Commercial Applications*. Chichester: Wiley.
12 Viterbi, A.J. (1995). *CDMA: Principles of Spread Spectrum Communication, Addison-Wesley Wireless Communications Series*. Reading, MA: Addison-Wesley.
13 Goodman, D.J. (1997). *Wireless Personal Communication Systems*. Reading, MA: Addison-Wesley.
14 Schiller, J. (2003). *Mobile Communications, Addison-Wesley Wireless Communications Series*, 2e. Reading, MA: Addison-Wesley.

15 Molisch, A.F. (2007). *Wireless Communications*. Chichester: Wiley.

16 Blaunstein, N. and Christodoulou, C. (2007). *Radio Propagation and Adaptive Antennas for Wireless Communication Links*, 1e. Hoboken, NJ: Wiley.

17 Blaunstein, N. and Christodoulou, C. (2014). *Radio Propagation and Adaptive Antennas for Wireless Communication Networks – Terrestrial, Atmospheric and Ionospheric*, 2e. Hoboken, NJ: Wiley.

8

MIMO Modern Networks Design in Space and Time Domains

Before starting to introduce the reader to the advanced multiple-input-multiple-output (MIMO) technology based on the usage of multibeam-adaptive antennas (usually called smart antennas in the literature, see bibliography in [1–17]), let us formulate the main goal of its adaptation for wireless networks beyond 3G. As was mentioned in [18–65], the MIMO systems are forecasted to provide higher data rates in a limited bandwidth by enabling the efficient use of spectrum. The main goal of MIMO systems evolutionary design is schematically presented in Figure 8.1.

8.1 Main Principles of MIMO

The main principle of increased throughput using MIMO is based on the presence of decorrelation between the multiple replicas of the received or transmitted signal [48–53]. This higher decorrelation can be achieved in the rich scattering environment where the capacity is linearly increased with the number of minimum transmit and receive antennas, which has been proven by [1–5]. Generally, the decorrelation is defined by the function of antenna spacing, angle of arrival (AOA) information, Doppler, and time of arrival (TOA) characteristics. The effects of azimuth spread and time spread on MIMO capacity were studied in [19, 20]. It was shown that higher azimuth and delay spread channels are advantageous over flat-fading cases in terms of increasing capacity. However, the real scattering environment is different from scene to scene, and as a result the decorrelation characteristics always depend on the particular propagation conditions.

We now present some advanced technology concepts based on adaptive multibeam or phased-array antenna applications through the prism of the physical layer description accounting for the "reaction" of the multipath outdoor channel with fading on radio propagation within such a channel [16–32]. These techniques are fully described in Refs. [33–51].

Advanced Technologies and Wireless Networks Beyond 4G, First Edition.
Nathan Blaunstein and Yehuda Ben-Shimol.
© 2021 John Wiley & Sons, Inc. Published 2021 by John Wiley & Sons, Inc.

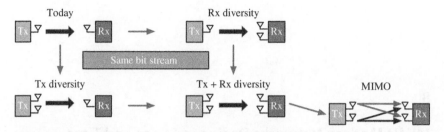

Figure 8.1 Evolutionary design – from Rx to Tx terminal antennas, through Tx and Rx diversity to MIMO system, by introducing more element antennas at both terminals

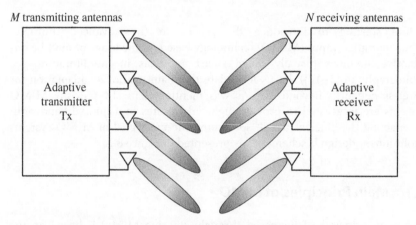

Figure 8.2 MIMO system based on $M \times N$ multibeam antennas

MIMO communication systems with multibeam or multiple-element antennas arranged at both ends of the communication link have been introduced during the past decade to increase spectral efficiency and communication link reliability that can be achieved via spatial and time diversity techniques. Figure 8.2 shows an example of how to arrange multibeam antennas (4×4 beams, i.e. 16 beams) that manage and control many clients serviced by the mobile/stationary broadband internet or sensors networking.

In modern MIMO systems, the two basic techniques usually used [32–51] to mitigate multipath fading phenomena and increase efficiency of such networks are:

1. Spatial multiplexing as a space–time modulation technique.
2. Diversity modulation technique as a special case of a space–time modulation technique.

According to the first procedure, each transmitting antenna element sends to the receiver independent (e.g. noncorrelated) streams of signal data accounting for a strong multipath phenomenon occurring in each channel of the MIMO system with the Rayleigh fading [32, 33] caused by multiple reflection and diffraction occurring in the real wireless channels (see Figure 8.3). This idea of spatial multiplexing was first proposed in Ref. [18], which then was adapted in practice of MIMO systems deployment in Refs. [35, 36]. Initially, spatial multiplexing systems used narrowband channel for each antenna element of the MIMO system with a small delay spread (i.e. the large bandwidth of coherency, see previous section). In modern MIMO systems, spatial multiplexing was adapted for wideband channels too in conjunction with orthogonal frequency-division multiplexing (OFDM) modulation technique [37–41].

As was shown in [15–17], the fade parameter K_i is reversely proportional to the weight parameter of each antenna element $\tilde{w}_i$, and the corresponding weight matrix $\mathbf{W}^T$ forms the relations between the outputs V_m and inputs U_m of an $M \times N$ planar array of the beam former at both the terminals, the basestation (BS) and the mobile subscriber (MS) (see Figure 8.3)

$$\mathbf{U} = \mathbf{W}^T \cdot \mathbf{V} \quad \det \mathbf{W} \neq 0 \tag{8.1}$$

where

$$\mathbf{W} = \begin{pmatrix} \tilde{w}_{1,1} & \cdots & \tilde{w}_{1,n} \\ \vdots & & \vdots \\ \tilde{w}_{m,1} & \cdots & \tilde{w}_{m,n} \end{pmatrix} \tag{8.2}$$

For single input-multiple output (SIMO) variant, the 2D matrix in (8.2) is deduced into a 1D-case:

$$U = \mathbf{W}^T \cdot \mathbf{V} = (\tilde{w}_1, \dots, \tilde{w}_m) \cdot \mathbf{V} \tag{8.3}$$

and $\tilde{w}_m \sim K_m^{-1}$ defines a multiplicative noise-to-signal ratio.

In contrast to spatial multiplexing, the diversity modulation technique deals with *dependent* (e.g. correlated) streams of data from each transmit antenna element of the desired MIMO system.

Figure 8.3 Schematically presented effects of multiray phenomena occurring in the real wireless MIMO network that causes strong fading on the input/output signals and interference between each element of the multibeam antenna

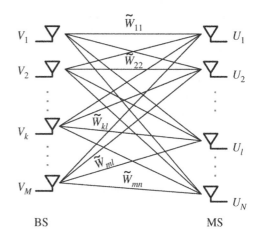

In this book, we use the term *MIMO diversity technique* that combines space–time diversity and spatial–time multiplexing, according to Refs. [47–65].

Following the results obtained in [17], we describe the MIMO system capacity, as an example of space- and time-diversity techniques adapted for the use of the multibeam (e.g. multielement) antennas in various built-up environments. For this case, the desired formulas according to the unified multiparametric stochastic approach described in [16, 17] (and discussed in Chapter 5) will be now rearranged to obtain simple relations between the MIMO antenna element number and the parameters of the terrain.

The spectral efficiency of the MIMO technique strongly depends on the diversity among multiple channels, which is determined by the spatial statistical behavior of the MIMO fading channel, partly characterized by the special spatial fading correlation function (or coefficients) described in [17]. It was shown in [17–24] that a sufficiently high diversity in the received multipath replicas of the transmitted signal can be achieved within a "rich" scattering environment where the communication channel capacity linearly increases with the number of the transmitter and the receiver antennas. Moreover, in [18] it was shown that the spatial fading correlation coefficient for *dependent* data streams (or decorrelation coefficient for *independent* data streams) can be determined by the system parameters, such as antenna elements spacing and their number. At the same time, this coefficient can be determined by the propagation conditions, such as the received energy spread in the AOA, TOA, and Doppler domains, as described in [16, 17].

In practice, scattering environment is scenario dependent, and as a result, spatial decorrelation characteristics are also scenario dependent [17–21, 24]. Therefore, an accurate modeling of the spatial decorrelation characteristics of the MIMO channels is crucial for the investigation of the scenario-specific spectral efficiency, which serves as the main criterion for communication systems design and wireless networks planning. Thus, we derive the MIMO channel capacity for various propagation conditions as a function of the spatial correlation and the received distribution in the AOA–TOA domain. The proposed spatial fading correlation will be introduced via the stochastic multiparametric model of the urban propagation conditions in a joint AOA–TOA domain described in [17–21, 24]. As a result of the proposed stochastic approach, it can be shown that the spatial fading correlation parameters depend on the propagation phenomena, such as multiple scattering, reflection, diffraction, and the waveguide propagation along streets, which characterize the urban environment propagation conditions.

8.2   Modeling of MIMO Channel Capacity

Usually, as follows from Figure 8.3, there are several output and input antennas assembled at the transmitter and the receiver, which will be denoted as M and N, respectively.

For the uncorrelated antennas (e.g. working as separate independent antenna elements) arranged at the MIMO channel, the capacity C, spectrally normalized to the bandwidth B_w (in Hz) (e.g. *spectral efficiency* measured in bits/s/Hz), was defined in [16–29] as

$$\tilde{C}_{\text{uncorr}} = N \log_2 \left[1 + \left(\frac{M}{N} \right) \left(\frac{K_m \left(\frac{P_m}{N} \right)_{\text{add}}}{K_m + \left(\frac{P_m}{N} \right)_{\text{add}}} \right) \right] \tag{8.4}$$

where $(P_m/N)_{\text{add}}$ is the signal-to-additive Gaussian noise ratio, which usually is taken into account in the literature. We also account the multiplicative noise caused by fading multipath phenomena, occurring in each of the N input channels, where $K_m = $ (LOS component)/(Multipath component) is the ratio of the coherent (e.g. deterministic) component of the signal and incoherent (stochastic) component of the signal that caused the multiplicative noise. At the same time, for the case of correlated antennas (i.e. working as unified whole transmitter and receiver antenna) the spectral efficiency, measured in bits/s/Hz, can be presented according to [16–19] as

$$\tilde{C}_{\text{corr}} = N \log_2 \left[1 + \left(\frac{MNK_m \left(\frac{P_m}{N} \right)_{\text{add}}}{K_m + \left(\frac{P_m}{N} \right)_{\text{add}}} \right) \right] \text{ (bits/s/Hz)} \tag{8.5}$$

Expression (8.5) can be verified by special numerical simulations carried out for different values of signal-to-noise ratio, SNR $= (P_m/N)_{\text{add}}$ (in dB), and for various amounts of elements M and N at the transmitter (output) and receiver (input) antennas, respectively. We should notice before presenting some results of numerical computations that the uncorrelated arrangement of the MIMO antenna elements at both the terminal sides allows to obtain much higher capacity, e.g. higher spectral efficiency of each M and N channels with respect to the case of the correlated arrangement of the antenna elements into the unique antenna. This can be clearly seen from formulas (8.4) and (8.5), where in (8.4) the number of elements at the receiver N is outside the logarithmic function, and C increases linearly with increase in elements N at the input of the receiver. At the same time, in (8.5), N and M are inside the logarithm, that is, capacity and the spectral efficiency increase logarithmically (i.e. nonlinearly and very slowly) with increase in the number of elements N and M. In [19, 20] are presented several variants of MIMO system arrangements to the proof mentioned above.

Therefore, we present here only the case of the uncorrelated arrangement of the antenna elements, M and N. Thus, Figure 8.4 presents a spectral efficiency vs. the fading factor K, as the ratio of the coherent and incoherent (multipath) components of the signal at the input of the multielement ($M = 2$) transmitter antenna and multielement ($N = 4$) receiving antenna, according to the scenario shown in Figure 8.3.

As clearly seen from the presented illustration, with increase of the coherent component (e.g. line-of-sight [LOS] component) of the income signals at

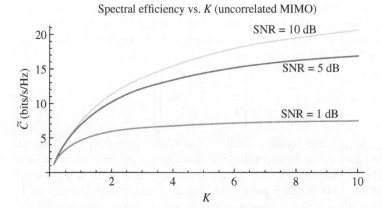

Figure 8.4 Spectral efficiency (in bits/s/Hz) vs. the K-fading parameter for various values of signal-to-additive noise ratio for $M = 2$ and $N = 4$

the input of the multielement receiver antenna with respect to the multipath component (caused by multidiffraction and multiscattering from obstructions located in area of users' service), that is, with increase of K-factor of fading, the spectral efficiency increases sharply till $K \sim 4$–6, and then a saturation of the process becomes evident.

This effect depends on signal-to-noise ratio (SNR) and becomes smaller ($K \sim 2$) with increase in SNR from 1 to 10 dB. In other words, increasing SNR inside the MIMO system, it can be easier to obey the effect of fading caused by real conditions of each land-to-land communication channel.

Considering now, according to [20], the uplink scenario where the multiple receiving antennas in the BS are spatially separated, the correlation among the received replicas of the transmitted signal is determined by the propagation conditions and the spatial separation distance, that is,

$$\rho = \int_0^{2\pi} e^{ikr\sin(\phi)} f(\phi) d\phi \tag{8.6}$$

where ϕ is AOA, $f(\phi)$ is an angular spectrum, $k = 2\pi/\lambda$ is the wavenumber, λ is the wavelength, and r is the spatial spacing between the transmitter and the receiver antennas, which in practice is limited by the physical dimensions of the platform or installation constraints. Conventionally, it is assumed that $M > N$.

The influence of the MIMO channel correlation, represented by the transmitter and the receiver spatial correlation matrices, on the channel capacity is intensively studied in the literature (see [17–20, 22]). Following [19, 20], we represent the influence of the spatial fading correlation in (8.4) on the MIMO channel capacity in (8.6) in the more convenient and simple manner to understand the matter:

$$\tilde{C} = N \log_2 \left(1 + (1 - \rho) \frac{M}{N} \frac{\text{SNR}}{B_w} \right) \tag{8.7}$$

Notice that the spatial fading correlation in (8.6) decreases the MIMO channel capacity. In addition, note that the received energy spread in the AOA–TOA domain is inversely proportional to the correlation and therefore directly proportional to the achievable channel capacity. Thus, the higher received signal spread is represented by a smaller correlation, $\rho \to 0$, which results in a higher MIMO channel capacity according to Eq. (8.7).

8.3 Fading Correlation in Space–Time Doman in Urban Environment with Dense Building Layout

Considering the below-the-rooftops propagation conditions [16, 17], and using the corresponding formulas presented in [20], the fading correlation in (8.6) can be rewritten as follows:

$$\rho_{\text{below}} = \int_0^{2\pi} f_{\text{below}}(\phi)e^{ikd\sin(\phi)}d\phi \tag{8.8}$$

where

$$f_{\text{below}}(\phi) = \int_{\tau=0}^{t} \left(\frac{\overline{L}v^2\tau d^2(\tau^2 - 1)}{2\pi(\tau - \cos\phi)} \beta(\phi)e^{-\frac{2\overline{L}v\tau d}{\pi}} \right) e^{\left(-\frac{d(\tau^2-1)|\ln\chi|}{(\tau - \cos\phi)a'(\phi)} \right)} d\tau \tag{8.9}$$

Here, d is the straight-line distance between the MS and the BS in km, τ is the ratio between the actual distance that signal travels from the MS to the BS and d, ϕ is the AOA, v is the building density in square kilometer, and

$$\beta(\phi) = \sin^2 \left(\frac{1}{2}\arcsin\left(\frac{d\sin(\phi)}{\tilde{r}} \right) \right) \tag{8.10}$$

$\chi = \overline{L}/(\overline{L} + \overline{l})$ is the street "discontinuity" parameter, where $\overline{L}$ is the average lengths of the buildings in km, $\overline{l}$ is the average lengths of the slits (gaps) between the buildings in km, $a' = \sqrt{\frac{4a^2}{\lambda^2 n^2} + a^2}$, where a is the average width of the streets in km, and n is the street waveguide mode number [17].

Considering the above-the-rooftops wave propagation [16, 17], and using formulas derived in [20], the correlation coefficient in (8.6) can be obtained as

$$\rho_{\text{above}} = \int_0^{2\pi} f_{\text{above}}(\phi)e^{ikd\sin\phi}d\phi \tag{8.11}$$

where

$$f_{\text{above}}(\phi) = \int_{\tau=0}^{t} \left[\left(\frac{\overline{L}v^2\tau d^2(\tau^2 - 1)}{2\pi(\tau - \cos\phi)} \beta(\phi)\frac{\overline{h}}{h_R}e^{-\frac{2\overline{L}v}{\pi}\left(\tilde{r}(\tau,\phi) + \frac{\overline{h}}{h_R}\frac{d(\tau^2-1)}{2(\tau - \cos\phi)} \right)} \right) + \right.$$
$$\left. \frac{v}{2}\beta(\phi)\frac{(h_R - \overline{h})\tilde{r}(\tau,\phi)}{\overline{h}}e^{\frac{2\overline{L}v\tilde{r}(\tau,\phi)}{\pi}} \right] e^{-\frac{d(\tau^2-1)|\ln\chi|}{(\tau - \cos\phi)a'(\tau,\phi)}} d\tau \tag{8.12}$$

Here, assuming a uniform distribution of building heights, we get, according to [16, 17], the average height of building profiles, $\bar{h} = (h_1 + h_2)/2$, where h_1 and h_2 are the minimum and maximum building heights, and h_R is the BS antenna height.

According to [16, 17], a radio path of the radio signal from MS antenna, as the receiver, to the BS antenna, as the transmitter, after scattering from any obstruction can be presented via the real path r of LOS, the distance between BS and MS, d, and the angle between the arriving scattered signal and the direct line between MS and BS antennas, that is, as

$$\tilde{r} = \sqrt{d^2 + r^2 - 2rd\cos\phi}$$

On the other hand, this variable can be also presented via the time delay between the direct signal from BS and the signal scattered from obstruction, as well as via the same scattered angle, that is, as

$$\tilde{r}(\tau, \phi) = \frac{d(\tau^2 - 2\tau\cos\phi + 1)}{2(\tau - \cos\phi)}$$

See [16, 17] and also Chapter 5. In the following section, we analyze the capacity of a MIMO channel for a specific urban environment.

8.4 Correlation Coefficient Analysis in Urban Scene

The spatial fading correlation is analyzed in various urban propagation conditions simulated by the proposed f_{below} and f_{above} models, according to expressions (8.9) and (8.12), respectively. We analyze an urban environment with the parameters of the experiment described in [20]: the two BS receiving antennas with a separation distance of $r/\lambda = 10$ are located in the urban scene with the following parameters: $\gamma_0 = 8$ km^{-1}, $\chi = 0.5$, $d = 0.3$ km, pseudo-LOS is at $\phi = 0°$, $\bar{h} = 15$ m, and $v = 250$ km^2. Figure 8.5 shows the correlation coefficient ρ from (8.11) as a function of the BS antenna height h_R and the buildings' density parameter, $v = 2\pi\bar{L}\nu/\pi$, which describes the clutter density in urban scenario.

As follows from the results presented in Figure 8.5, the spatial fading correlation is directly proportional to the BS antenna height. We can outline that the results shown in Figure 8.5 agree with those obtained in [21, 24]. In addition, Figure 8.5 shows that the spatial fading correlation is inversely proportional to the buildings' density parameter, γ_0 (e.g. clutter density), which determines the received signal diversity in the AOA–TOA domain.

Therefore, the denser urban environment (higher γ_0) results in lower correlation among the received replica of the transmitted signal. Finally, we notice that the influence of the parameter γ_0 on the spatial fading correlation decreases with increasing BS antenna height.

This observation can be explained by the fact that the significance of the built-up environment structure for a particular urban scene (distribution of

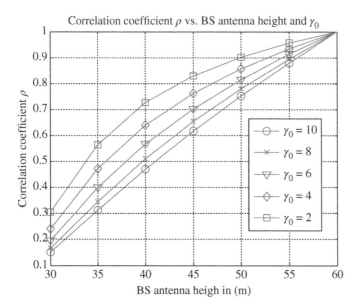

Figure 8.5 Correlation coefficient vs. the BS antenna height h_R for various $\gamma_0 = 2, 4, 6, 8, 10$ km^{-1} in an urban environment with $r = 10\lambda$, $\chi = 0.5$, $d = 0.3$ km, pseudo-LOS is at $0°$, $\bar{h} = 15$ m, and $v = 250$ per km^2

the scatterers) decreases with increasing BS antenna height. In practice, the result shown in Figure 8.5 can be used as a guideline for the BS antenna height selection required to achieve the predefined correlation (and as a result the predefined MIMO channel capacity) in a specific urban scenario.

8.5 MIMO Channel Capacity Estimation

Here, following Ref. [65] we analyze the effect of the spatial fading correlation on the MIMO channel capacity using (8.6) and (8.10) in (8.12). Figure 8.6a shows the MIMO channel capacity as a function of the BS antenna heights for a variety of the buildings' density values (i.e. γ_0) in the simulated urban environment with the following parameters: SNR = 10 dB, separation distance between two BS receiver antennas of $r = 10\lambda$, $\chi = 0.5$, $d = 0.5$ km, pseudo LOS is at $\phi = 0°$, and $\bar{h} = 25$ m.

Figure 8.6a shows that the MIMO channel capacity is inversely proportional to the spatial fading correlation shown in subplot (b). Thus, the increase in the spatial fading correlation ρ from 0.2 to 0.8 results in the normalized channel capacity degradation from approximately 6.5 to 3.2 bits/s/Hz. Notice that the MIMO channel capacity is directly proportional to the buildings' density γ_0 in

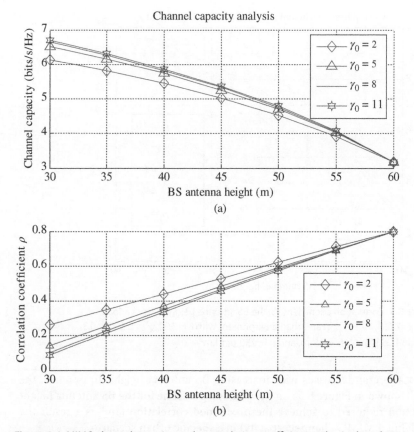

Figure 8.6 MIMO channel capacity vs. the correlation coefficient and γ_0 in the urban scenario with SNR = 10 dB, $r = 10\lambda$, $\chi = 0.5$, $d = 0.5$ km, pseudo LOS is at $\phi = 0°$, and $\bar{h} = 25$ m

the urban environment. Thus, the increase in the density γ_0 from 2 to 11 km^{-1} results in channel capacity improvement.

These simulations show that in addition to the conventional dependence of the MIMO channel capacity on SNR, it strongly depends on the location-specific parameters of the urban environment (via the statistical model of the urban propagation conditions), such as the BS antenna height, buildings' density, and the average buildings' height.

8.6 Analysis of MIMO Channel Capacity in Predefined Urban Scenario

Now we evaluate the MIMO channel capacity (or the spectral efficiency, as the capacity-to-frequency band ratio, in bits/s/Hz) in a simulated "virtual"

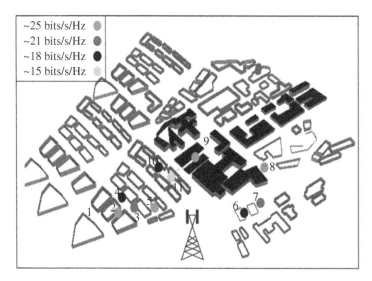

~25 bits/s/Hz
~21 bits/s/Hz
~18 bits/s/Hz
~15 bits/s/Hz

Figure 8.7 The MIMO channel capacity in the simulated urban scenario with the BS antenna height of 15 m, MIMO system with 4 × 4 antenna elements, and SNR = 20 dB; the numbers show places where the numerical experiment was carried out

urban scenario, using an urban topographic plan taken from [17, 20] (the same topographic plan of Helsinki was analyzed in Chapter 5, Scenario 1). The MIMO channel capacity was evaluated in 11 representative locations characterized by the SNR of 20 dB. The SNR was obtained using the ray-tracing software tool [23] as a ratio between the received signal strength and the noise with an equivalent bandwidth of 5 MHz. A BS antenna height of 15 m was simulated. The separation distance of $r = 10\lambda$ was simulated between four receiving antennas. The urban environment with the following parameters was simulated using (8.11) and (8.12): $\chi = 0.8$, $\bar{h} = 15$ m, and $\gamma_0 = 6$ km^{-1}. The interuser interference was neglected in this simulation.

Figure 8.7 shows the simulated urban scene. The dots in Figure 8.7 show the tested MS locations with the SNR of 20 dB. The pattern represents the achievable 4 × 4 MIMO channel capacity. Notice that different locations (with equal SNR) are characterized by different spatial fading correlations, and therefore the resulting MIMO channel capacity varies among these locations. We also notice that results shown in Figure 8.7 agree with the results obtained during the measurement reported in Ref. [17].

The above obtained results allow us to summarize that the MIMO channel capacity in an urban environment has the BS antenna elevation and location-specific factor and strongly depends on the spatial fading correlation determined by the scattering, reflection, diffraction, and waveguide propagation phenomena.

References

1 Steele, R. (1992). *Mobile Radio Communication*. IEEE Press.

2 Proakis, J.G. (1995). *Digital Communications*, 3e. New York: McGraw-Hill.

3 Stuber, G.L. (1996). *Principles of Mobile Communications*. Boston, MA: Kluwer Academic Publishers.

4 Rappaport, T.S. (1996). *Wireless Communications: Principles and Practice*, 2e in 2001. New York: Prentice Hall PTR.

5 Steele, R. and Hanzo, L. (1999). *Mobile Communications*, 2e. Chichester: Wiley.

6 Li, J.S. and Miller, L.E. (1998). *CDMA Systems Engineering Handbook*. Boston, MA and London: Artech House.

7 Burr, A. (2001). *Modulation and Coding for Wireless Communications*. New York: Prentice Hall PTR.

8 Paetzold, M. (2002). *Mobile Fading Channels: Modeling, Analysis, and Simulation*. Chichester: Wiley.

9 Simon, M.K., Omura, J.K., Scholtz, R.A., and Levitt, B.K. (1994). *Spread Spectrum Communications Handbook*. New York: McGraw-Hill.

10 Glisic, S. and Vucetic, B. (1997). *Spread Spectrum CDMA Systems for Wireless Communications*. Boston, MA and London: Artech House.

11 Dixon, R.C. (1994). *Spread Spectrum Systems with Commercial Applications*. Chichester: Wiley.

12 Viterbi, A.J. (1995). *CDMA: Principles of Spread Spectrum Communication, Addison-Wesley Wireless Communications Series*. Reading, MA: Addison-Wesley.

13 Goodman, D.J. (1997). *Wireless Personal Communication Systems*. Reading, MA: Addison-Wesley.

14 Schiller, J. (2003). *Mobile Communications, Addison-Wesley Wireless Communications Series*, 2e. Reading, MA: Addison-Wesley.

15 Molisch, A.F. (2007). *Wireless Communications*. Chichester: Wiley.

16 Blaunstein, N. and Christodoulou, C. (2007). *Radio Propagation and Adaptive Antennas for Wireless Communication Links*, 1e. Hoboken, NJ: Wiley.

17 Blaunstein, N. and Christodoulou, C. (2014). *Radio Propagation and Adaptive Antennas for Wireless Communication Networks – Terrestrial, Atmospheric and Ionospheric*, 2e. Hoboken, NJ: Wiley.

18 Andersen, J.B. (2000). Array gain and capacity of known random channels with multiple element arrays at both ends. *IEEE J. Sel. Areas Commun.* 18:2172–2178.

19 Blaunstein, N. and Yarkoni, N. (2006). Capacity and spectral efficiency of MIMO wireless systems in multipath urban environment with fading. *Proceedings of the European Conference on Antennas and Propagation, EuCAP-2006, Nice, France, pp. 111–115.*

20 Tsalolihin, E., Bilik, I., and Blaunstein, N. (2011). MIMO capacity in space and time domain for various urban environments. *Proceedings of 5th European Conference on Antennas and Propagation* (EuCAP), Rome, Italy (11–15 April 2011), pp. 2321–2325.

21 Molisch, A., Steinbauer, M., Toeltsch, M. et al. (2002). Capacity of MIMO systems based on measured wireless channels. *IEEE J. Sel. Areas Commun.* 20:561–569.

22 Gesbert, D., Shafi, M., Shiu, D. et al. (2003). From theory to practice: an overview of MIMO space-time coded wireless systems. *IEEE J. Sel. Areas Commun.* 21 (3):281–302.

23 Shiu, M.D., Foschini, G.J., and Kahm, J.M. (2000). Fading correlation and its effect on the capacity of multi-antenna systems. *IEEE Trans. Commun.* 48:502–513.

24 Philippe, J., Schumacher, L., Pedersen, K. et al. (2002). A stochastic MIMO radio channel model with experimental validation. *IEEE J. Sel. Areas Commun.* 20 (6):1211–1226.

25 Gesbert, D., Boleskei, H., Gore, D.A., and Paulraj, A.J. (2002). Outdoor MIMO wireless channels: models and performance prediction. *IEEE Trans. Commun.* 50 (6):1926–1934.

26 Boleskei, H., Borgmann, M., and Paulraj, A.J. (2002). On the capacity of OFDM-based spatial multiplexing systems. *IEEE Trans. Commun.* 50 (1):225–234.

27 Boleskei, H., Borgmann, M., and Paulraj, A.J. (2003). Impact of the propagation environment on the performance of space-frequency coded MIMO-OFDM. *IEEE J. Sel. Areas Commun.* 21 (2):427–439.

28 Chizik, D., Ling, J., Wolniansky, P.W. et al. (2003). Multiple-input-multiple-output measurements and modeling in Manhattan. *IEEE J. Sel. Areas Commun.* 23 (2):321–331.

29 Oyman, O., Nabar, R.U., Boleskei, H., and Paulraj, A.J. (2003). Characterizing the statistical properties of mutual information in MIMO channels. *IEEE Trans. Signal Process.* 51:2784–2795.

30 Paulraj, A.J., Gore, D.A., Nabar, R.U., and Boleskei, H. (2004). An overview of MIMO communications – a key to gigabit wireless. *Proc. IEEE* 92 (2):198–218.

31 Forenza, A., McKay, M.R., Pandharipande, A. et al. (2007). Adaptive MIMO transmission for exploiting the capacity of spatially correlated channels. *IEEE Trans. Veh. Technol.* 56 (2): 619–630.

32 Foschini, G.J. and Gans, M.J. (1998). On limits of wireless communications in a fading environment when using multiple antennas. *Wireless Person. Commun.* 6 (3):311–335.

33 Proakis, J.G. (2001). *Digital Communications*, 4e. New York: McGraw-Hill.

34 Paulraj, A.J. and Kailath, T. (1994). Increasing capacity in wireless broadcast systems using distributed transmission/directional reception (DTDR). US Patent 5, 345, 599, 6 September 1994.

35 Foschini, G.J. (1996). Layered space-time architecture for wireless communication in a fading environment when using multiple antennas. *Bell Labs. Tech. J.* 1 (2):41–59.

36 Golden, G.D., Foschini, G.J., Valenzula, R.A., and Wolniansky, P.W. (1999). Direction algorithm and initial laboratory results using the V-BLAST space-time communication architecture. *Electron. Lett.* 35 (1):14–15.

37 Nabar, R.U., Bolcskei, H., Erceg, V. et al. (2002). Performance of multi-antenna signaling techniques in the presence of polarization diversity. *IEEE Trans. Signal Process.* 50 (10):2553–2562.

38 Zheng, L. and Tse, D. (2003). Diversity and multiplexing: a fundamental tradeoff in multiple antenna channels. *IEEE Trans. Inform. Theory* 49 (5):1073–1096.

39 Varadarajan, B. and Barry, J.R. (2002). The rate-diversity trade-off for linear space-time codes. *Proc. IEEE Veh. Tech. Conf.* 1:67–71.

40 Godovarti, M. and Nero, A.O. (2002). Diversity and degrees of freedom in wireless communications. *Proc. ICASSP* 3:2861–2864.

41 Raleigh, G.G. and Cioffi, J.M. (1998). Spatio-temporal coding for wireless communication. *IEEE Trans. Commun.* 46 (3):357–366.

42 Wittniben, A. (1991). Base station modulation diversity for digital simulcast. *Proc. IEEE Veh. Tech. Conf.* 848–853.

43 Seshadri, N. and Winters, J.H. (1994). Two signaling schemes for improving the error performance of frequency-division-duplex (FDD) transmission systems using transmitter antenna diversity. *Int. J. Wireless Inform. Netw.* 1 (1):49–60.

44 Alamouti, S.M. (1998). A simple transmit diversity technique for wireless communications. *IEEE J. Sel. Areas Commun.* 16 (8):1451–1458.

45 Tarokh, V., Seshandri, N., and Calderbank, A.R. (1999). Space-time codes for high data rate wireless communication: performance criterion and code construction. *IEEE Trans. Inform. Theory* 45 (5):1456–1467.

46 Ganesan, G. and Stoica, P. (2001). Space-time block codes: a maximum SNR approach. *IEEE Trans. Inform. Theory* 47 (4):1650–1656.

47 Hassibi, B. and Hochwald, B.M. (2002). High-rate codes that are linear in space and time. *IEEE Trans. Inform. Theory* 48 (7):1804–1824.

48 Health, R.W. Jr. and Paulraj, A.J. (2002). Linear dispersion codes for MIMO systems based on frame theory. *IEEE Trans. Signal Process.* 50 (10):2429–2441.

49 Winters, J.H. (1998). The diversity gain of transmit diversity in wireless systems with Rayleigh fading. *IEEE Trans. Veh. Technol.* 47 (1):119–123.

50 Bjerke, B.A. and Proakis, J.G. (1999). Multiple-antenna diversity techniques for transmission over fading channels. *Proc. Wireless Commun. Netw. Conf.* 3:1038–1042.

51 3GPP Technical Specification Group Radio Access Network (2010). Evolved Universal Terrestrial Radio Access (E-UTRA), Physical Layer Procedures (Release 9). 3GPP TS36.213 V9.3.0, June 2010.

52 3GPP Technical Specification Group Radio Access Network (2010). Evolved Universal Terrestrial Radio Access (E-UTRA), Further advancements for E-UTRA physical layer aspects (Release 9). 3GPPTS36.814V9.0.0, March 2010.

53 Ghaffar, R. and Knopp, R. (2011). Interference-aware receiver structure for Multi-User MIMO and LTE. *EURASIP J. Wireless Commun. Network.* 40:24.

54 Li, Q., Li, G., Lee, W. et al. (2010). MIMO techniques in WiMAX and LTE: a future survey. *IEEE Commun. Mag.* 48 (5): 86–92.

55 Kusume, K., Dietl, G., Abe, T. et al. (2010). System level performance of downlink MU-MIMO transmission for 3GPP LTE-advanced. *Proceedings of the IEEE Vehicular Technology Conference-Spring (VTC '05)*, Ottawa, Canada (September 2010), 5 pages.

56 Covavacs, I.Z., Ordonez, L.G., Navarro, M. et al. (2010). Toward a reconfigurable MIMO downlink air interface and radio resource management: the SURFACE concept. *IEEE Commun. Mag.* 48 (6): 22–29.

57 EU FP7 Project SAMURAI – Spectrum Aggression and Multi-User MIMO: Real-World Impact. http://www.ict-samurai.eu/page1001.en.htm (accessed 9 May 2020).

58 3GPP TSG RAN WG1 #62 (2010). Way Forward on Transmission Mode and DCI design for Rel-10 Enhanced Multiple Antenna Transmission. R1-105057, Madrid, Spain, August 2010.

59 3GPP TSG RAN WG1 #62 (2010). Way Forward on 8 Tx Codebook for Release 10 DL MIMO. R1-105011, Madrid, Spain, August 2010.

60 3GPP TR 36.942 V10.3.0 (2012-06), 3rd Generation Partnership Project (2012). Technical Specification Group Radio Access Network; Evolved Universal Terrestrial Radio Access (E-UTRA); Radio Frequency (RF) system scenarios (Release 10), June 2012.

61 3G Americas white paper (2010). 3GPP mobile broadband innovation path to 4G: Release 9, Release 10 and beyond: HSPA+, SAE/LTE and LTE-advanced. http://www.4gamericas.org/documents/3GPP_Rel-9_Beyond%20Feb%202010.pdf (accessed 9 May 2010).

62 Duplicy, J., Badic, B., Balraj, R. et al. (2011). MU-MIMO in LTE Systems. *EURASIP J. Wireless Commun. Netw.*Article ID 496763, 13 pages. https://doi.org/10.1155/2011/496763.

63 3GPP TR 36.942 V8.4.0 (2012-06), 3rd Generation Partnership Project (2012). Technical Specification Group Radio Access Network; Evolved Universal Terrestrial Radio Access (E-UTRA); Radio Frequency (RF) system scenarios (Release 8), June 2012.

64 Zhang, H., Prasad, N., and Rangarajan, S. (2011). MIMO Downlink Scheduling in LTE and LTE-Advanced Systems. *Tech. Rep.* NEC Labs America. http://www.nec-labs.com/honghai/TR/lte-scheduling.pdf (accessed 9 May 2020).

65 Blaunstein, N.Sh. and Sergeev, M.B.S. (2012). Integration of advanced LTE technology and MIMO network based on adaptive multi-beam antennas. In: *Internet of Things, Smart Spaces, and Next Generation Networking*(ed. S. Andreev, S. Balandin, and Y. Koucheryavy) (Proceedings of 12th International Conference on NEW2AN 2012 and 5th Conference ruSMART 2012, St. Petersburg, Russia (27–29 August 2012)), 164–173. Heidelberg, Dordrecht, London, and New York: Springer-Verlag.

9

MIMO Network Based on Adaptive Multibeam Antennas Integrated with Modern LTE Releases

9.1 Problems in LTE Releases Deployment

It should be outlined that even using a single-carrier (SC) frequency-division multiple access (FDMA) modulation technique for uplink transmission and an SC-orthogonal frequency-division multiple access (OFDMA) technique for downlink transmissions (see definitions in [1]), it is difficult to provide a wide range of spectra allocations of different sizes for each subscriber located in various terrestrial conditions, as well as a significant increase in spectrum efficiency compared to previous 2G and 3G cellular networks. This can be achieved only by combining advanced FDMA and OFDMA technologies with multiple-input-multiple-output (MIMO) systems performed on the basis of multibeam or phased-array antennas [2–13]. The long-term evolution (LTE) Release 8 was recently expanded from two to four antennas in downlink spatial multiplexing from a base station (BS), as shown in Figure 9.1 (called also SIMO-LTE system).

Here, the layers can be defined as simultaneously transmitted streams of data to multiple user equipments (UEs) using the same time–frequency resource. In such a manner, any transmission of separate data streams is distributed among desired layers. The LTE Release 9, as a new dual-layer transmission mode, was also performed for supporting of up to four transmitted antennas at the BS in downlink channel. Now we postulate the following question: If both LTE Releases 8 and 9 could be integrated with MIMO, can such a combined LTE-MIMO system satisfy the International Mobile Telephony (IMT) requirements? Table 9.1 presents a comparative analysis of these requirements with respect to those that satisfy the deployments of LTE Release 8 system. It is clearly seen that, even giving better latency for each EU, but using twice-narrower bandwidth, Release 8 cannot support high-rate transmission of data streams for each individual user. A fully adaptive multiuser (MU) MIMO transmission mode cannot be realized in cooperation with LTE Releases 8 and 9 [2, 3, 12, 14].

Advanced Technologies and Wireless Networks Beyond 4G, First Edition.
Nathan Blaunstein and Yehuda Ben-Shimol.
© 2021 John Wiley & Sons, Inc. Published 2021 by John Wiley & Sons, Inc.

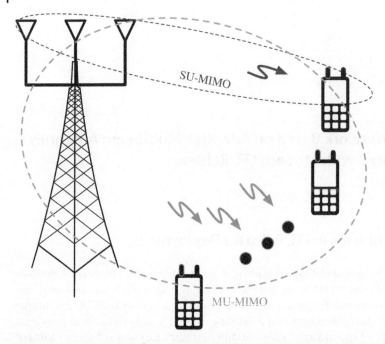

Figure 9.1 Geometrical configuration of SU-MIMO and MU-MIMO in integrated LTE-MIMO system

Table 9.1 Requirements of IMT-advanced (LTE Release 10) vs. LTE Release 8 system (extracted from [2, 3])

		IMT: advanced requirement	LTE release 8
Transmission bandwidth		At least40 MHz	Up to20 MHz
Peak spectral efficiency	Downlink (DL)	15 bps/Hz	16 bps/Hz
	Uplink (UL)	6.75 bps/Hz	4 bps/Hz
Latency	Control plane	< 100 ms	50 ms
	User plane	< 10 ms	4.9 ms

Recently, a new MU MIMO antenna system was introduced called the Advanced LTE (A-LTE) or, simply, LTE Release 10 [10–12]. We introduce this advanced technology in Table 9.1 since, as was mentioned in [15, 16], it is better equipped to meet the requirements of the modern fourth (4G) and fifth (5G) generations of wireless networks.

As seen in Figure 9.2, the LTE Release 10, or A-LTE, can be used at BS with at least eight separate antennas for downlink MU connections, whereas for

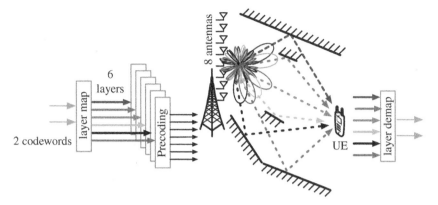

Figure 9.2 Downlink transmission from BS arranged by eight antennas to users in MIMO-LTE-integrated network (rearranged from [10, 11])

uplink, up to four UE antennas can be utilized. Here, a layer mapping supports the transfer of individual codes from two codebooks to each predecoding layer.

At the MU terminals, a demapping layer is used for transporting to each individual user its desired data codes. The use of a MIMO system at both end terminals allows for:

- fast user channel estimation, selecting, and equalization;
- reliable cancellation of MU interference;
- simplification of the complexity of the interference-aware (IA) receiver;
- reduction of the system's detection complexity;
- fitting of each single antenna of UEs in MIMO configuration;
- better implementation in the existing hardware, and so on.

In Refs. [13, 15] the scheduling algorithms, based on the geometrical alignment at the BS, which can minimize the interuser interference (IUI) seen by each UE, were introduced. In such a configuration, the proposed IA receiver was found as a good candidate for the practical implementation in MU-MIMO LTE Release 10 combined system.

To show the difference between the single user–single input–single output (SU-SISO) and MU-MIMO systems based on multibeam antennas, we schematically presented them in Figure 9.3a,b. The first system is based on point-to-point single BS and single UE antennas, whereas the second one is based on multiple antennas from both terminals [16].

9.2 Multibeam MIMO with Adaptive Antennas Against Fading Phenomena in LTE Networks

The main goal of the usage of multibeam antennas was to deploy in the existing and advanced communication networks the so-called spatial filtering of each

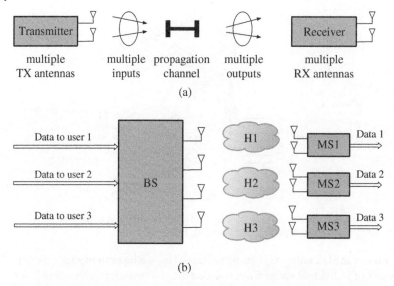

Figure 9.3 (a) SU-MIMO network and (b)MU-MIMO network based on multibeam antennas

desired signal. Usage of multiple beams concept with narrow beamwidth allows eliminating the IUI, minimizing the multiplicative noise caused by fading phenomena occurring in the multipath communication channels, mostly for urban and suburban environments, and finally, tracking each mobile subscriber (MS) during his existence in the area of service of multibeam BS antenna. Unfortunately, as was shown in [1], in each direction in azimuth and elevation domains for a current position of the desired subscriber the channel specific response should be introduced, and it was proposed to define this response by *Ricean K*-parameter of fading, as a ratio of the coherent (e.g. signal) and incoherent (e.g. noisy) components of the recorded signal. This parameter was evaluated for different terrestrial scenarios, rural, mixed-residential, suburban, and urban, and was shown that this parameter fully depends on the features of the built-up terrain, such as the buildings' density, overlay profile, orientation, and elevation with respect to the BS and MS antenna, heights of the terminal antennas, and so on.

Thus, for the mixed-residential area using the results obtained in [1], we get the expression for K-factor along the radio path d between two terminal antennas:

$$K = \frac{I_{co}}{I_{inc}} = \frac{\exp\left(-\gamma_0 d \frac{\bar{h}-z_1}{z_2-z_1}\right)\left[\frac{\sin(kz_1z_2/d)}{2\pi d}\right]^2}{\frac{\Gamma}{8\pi}\frac{\lambda\ell_h}{\lambda^2+(2\pi\ell_h\gamma_0)^2}\frac{\lambda\ell_v}{\lambda^2+[2\pi\ell_v\gamma_0(\bar{h}-z_1)]^2}\frac{\sqrt{(\lambda d/4\pi^3)+(z_2-\bar{h})^2}}{d^3}} \quad (9.1)$$

For the urban and suburban environments following [17], we finally get

$$K = \frac{I_{co}}{I_{inc1} + I_{inc2}} =$$

$$\frac{\exp\left(-\gamma_0 d \frac{z_1 - \bar{h}}{z_2 - z_1}\right) \frac{\sin^2(k z_1 z_2/d)}{4\pi^2 d^2}}{\frac{\Gamma \lambda \ell_v \sqrt{\lambda d/4\pi^3 + (z_2 - \bar{h})^2}}{8\pi[\lambda^2 + (2\pi \ell_v \gamma_0 (z_1 - \bar{h})^2)]d^3} + \frac{\Gamma^2 \lambda^3 \ell_v^2 [\lambda d/4\pi^3 + (z_2 - \bar{h})^2]}{24\pi^2[\lambda^2 + (2\pi \ell_v \gamma_0 (z_1 - \bar{h}))^2]^2 d^3}}
\tag{9.2}$$

Here, as above, in expressions (9.1) and (9.2), as in Chapters 3 and 5, $\gamma_0 = 2\bar{L}v/\pi$ is the density of the buildings' contours (in km^{-1}), v is the density of the buildings in the area of service (in km^{-2}), $\bar{L}$ is the average length (or width) of the buildings (in m), depending on its orientation with respect to terminal antenna, Γ is the absolute value of the reflection coefficient, ℓ_v and ℓ_h are the vertical and horizontal scales of coherency of the reflections from the building walls, respectively (in m), $\bar{h}$ is the average buildings' height (in m), and z_1 and z_2 are the heights of the MS and BS antenna, respectively.

9.3 Analysis of the Multibeam Effect for a Specific Environment

In this section, we present some numerical experiments carried out for one of the practical urban environments of service, such as the microcellular area in Ramat-Gan built-up area shown in Figure 9.4 according to [1] (the same area was analyzed in Chapter 3 for the performance of the total path loss and link budget design).

The BS antenna height was $z_2 = 50$ m. The outdoor desired MSs were located on the ground at the height of $z_1 = 2$ m. The simulations were performed for the following parameters of propagation: $\bar{h} = 25$ m, $\gamma_0 = 10$ km^{-1}, and $\ell_v, \ell_h = 1 - 2$ m [1]. The results of simulation of the K-parameter of fading are shown in Figure 9.5a,b for the carrier frequency of $f = 1800$ and 2400 MHz, respectively.

Based on these results, we can now for each MS with number i estimate the signal-to-multiplicative noise ratio via K-factor that determines the losses and fading effects in each noisy communication channel for each desired UE. This procedure allows obtaining the channel spectral efficiency as a function of K for various configurations of MIMO system consisting M outputs and N inputs, by use of the corresponding formulas (8.4) and (8.5) for practical arrangement of multibeam antennas: uncorrelated and correlated, respectively.

To show the efficiency of usage of the combination of MIMO/LTE network based on multibeam antennas with respect to SISO network, we present the computations of a normalized maximum sum rate I (in bits/s/Hz) of downlink

Figure 9.4 Dense built-up area; numbers of building floors are indicated by the corresponding colors

based on the mathematical algorithm fully described in [4, 16]. In simulations, we account for an SU-SISO (that is, for $M = 1$, $N = 1$), for an SU-SIMO (that is, for $M = 1$, $N = 4$), and for an MU-MIMO (that is, for $M = 4$, $N = 1$) integrated schemes (in the case of NINO antenna with uncorrelated elements, i.e. using formula (8.4)). The results of the numerical experiment are shown in Figure 9.6.

It is clearly seen from the presented illustrations that using the MU-MIMO system of various input–output antenna elements integrated as an example, with an A-LTE technology, it is possible to increase the spectral efficiency and the data rate in such an integrated MU-MIMO network. Moreover, both SU-SIMO (or MISO) LTE and MU-MIMO LTE-integrated systems, with a high relation between transmitter and receiver multibeam antennas (see Figure 9.6), show better performance in spectral efficiency and data stream rate [16]. Thus, it can be seen that the LTE-SU Rx gives low spectral efficiency and data rate with respect to the MU-MIMO LTE-A system. The latter has a tendency to increase spectral efficiency and data rate per several times with respect to the previous systems, and this tendency increases with an increase

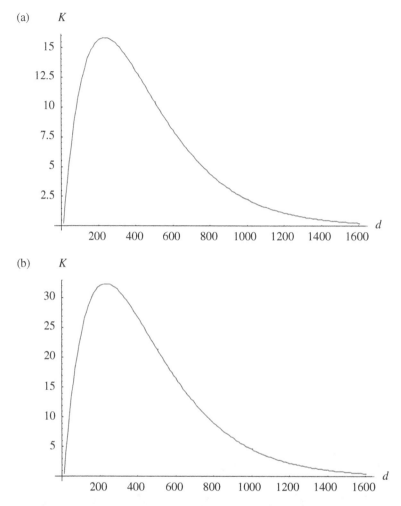

Figure 9.5 (a) The Ricean K-factor vs. the distance between BS and MSs in dense urban area for $f = 1.8$ GHz. (b) The same as in (a), but for $f = 2.4$ GHz

in SNR. With increase in the amount of UE antennas and BS antennas, this difference becomes more significant.

Such configurations can be extended for the combined femtocell/picocell/ microcell/macrocell planning tool design (see Figure 9.7). Finishing this paragraph, we should outline that by controlling the number of elements of multibeam antennas at the both end terminals, BS and UEs, and a priori accounting for the real responses of each channel on multipath fading phenomena (by prediction of the real K-factor of fading), we show the same effects, as were obtained in [10, 12–15], where a set of precoding codebooks (from one to several) was introduced for the extension of the LTE Releases

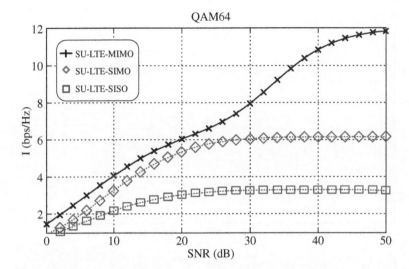

Figure 9.6 Spectral efficiency in bps/Hz: indicates the SU-LTE-SISO ($M = N = 1$) system, the SU-LTE-SIMO ($M = 4, N = 1$) system, and the MU-LTE-MIMO ($M = 4, N = 4$) system with uncorrelated antenna elements (using the same notations as in [4])

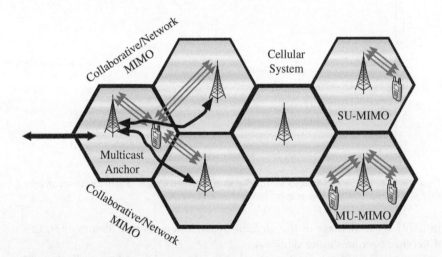

Figure 9.7 The proposed implementation of cellular, LTE, and SU/MU-MIMO systems into the integrated configuration of the current fourth generation and future fifth generations of the wireless networks (extracted from the internet)

8–10, using MIMO configurations with two or four transmitting antennas, or a dual-codebook deployment [4] for MIMO configuration with eight transmitting antennas at the BS terminal.

What is important finally to notice is that the main limiting factors that can decrease the efficiency of the proposed MIMO system integrated, for example, with the A-LTE Releases from 8 to10, observed during numerical computations based on the real experiment carried in the built-up area, are the K-factor of fading, as a response of each individual communication channel on data transmission, and the number of antenna beams within each terminal of the system, BS and UE.

Despite the fact that the approaches, proposed in [4, 10, 12–15], can reduce the total LTE/MIMO system structure yielding a low complexity of signal processing against interuser interference, however, as was shown in [16], without accounting for the K-factor of fading (based on the topographic layouts of the built-up terrain), and for the effective configuration of multibeam antennas [16], it is complicated apriori to predict data parameters for each subscriber channel. All these, finally, can increase the efficiency of the

Figure 9.8 LTE-MIMO network configuration based on multibeam antennas for the combined femtocell/picocell/microcell/macrocell planning tool design

proposed A-LTE/MIMO network and control of its grade-of-service (GOS) and quality-of-service (QOS).

Finally, following the results obtained in Refs. [5–16], as well as the recommendations stated there, we propose the following configuration of the SU-MIMO and MU MIMO integrated with A-LTE, "hidden" into the conventional cellular-map scheme, as is shown in Figure 9.8.

These schemes can be considered as the most convenient "candidates" of the wireless links configurations that satisfied the modern requirements of the fourth- and fifth-generation networks.

9.4 Summary

In Section 9.2, we introduced the reader to the conventional and current techniques, technologies and systems adapted for second (2G) and third (3G) generations of wireless networks, and the advanced technologies and their corresponding protocols used to utilize modern networks beyond 3G, such as fourth and fifth generations. New generations, called fourth (4G) and fifth (5G), were introduced which are expected to be capable of providing wider bandwidth, higher data stream rates, greater interoperability accords, various communication protocols, without any collision between them, user's security, and noncollision communications between users, that is, to provide significant increase in GOS and QOS.

Thus, typical 2G standards as GSM (Global System for Mobile Communications) operated at 900–1900 MHz frequency band, which used TDMA/FDMA separate or combined digital modulation techniques, have not satisfied the high-data communication requirement. The Universal Mobile Telecommunications Systems (UMTS) standards that were related to 2.5G and 3G mobile systems dealt with higher voice capacity and higher speed digital data. The same parameters were expected for 3G communication networks such as WPAN (or Bluetooth), Wi-Fi (or WLAN), WiMAX; all are described briefly above. Unfortunately, even integrating and combining the existing 2G and 3G networks, technologies, and protocols was problematic to achieve 200 Mbps—1 Gbps data rate, multimedia (video and audio) applications, and terminal's heterogeneity related to significant decrease of the network costs and greater mobile signal availability in a "jungle of noises" caused by multipath fading.

For these reasons in the fourth and fifth generations, a physical layer was significantly broadened by serving a wide range of frequency bands. In our opinion, recently performed modern LTE and LTE-Advanced (LTE-A) networks, integrated with MIMO systems based on multiantenna (multibeam or phased-array) technology, can substantially improve 4G network spectrum efficiency providing three kinds of advantages with respect to the single-antenna LTE system:

Table 9.2 IMT requirements for fourth generation vs. the last LTE releases and WiMAX3G generation of networks

		IMT advanced requirements	LTE		WiMaX	
			Rel. 8	Rel. 10	1.0	2.0
Transmission bandwidth (MHz)		≥ 40	≤ 20	≤ 100	≤ 10	≤ 40
Peak spectral efficiency	DL(bps/Hz)	15	16	16	6.4	15
	UL(bps/Hz)	6.75	2.8	8.1	4	6.75
Latency(ms)	Controlplane	30	50	50	50	< 100
	User plane	< 10	10	20	4.9	4.9

- transmit time diversity,
- beamforming, and
- spatial multiplexing.

All these advantages are shown in Table 9.2. Moreover, using spatial multiplexing, the number of simultaneously transmitted data streams, as well as the beam pattern for each transmitted data stream, can be managed and controlled by the corresponding protocol to optimize the fourth and fifth networks' capacity.

Therefore, such an integration of a MIMO system with LTE-A technology allows us to avoid, in practice, all the drawbacks related to the previous generations described above. It also allows designers of modern fourth generation of wireless networks to improve their GOS and QOS, protecting against ISI and ICI caused by multipath fading phenomena, increase their frequency spectral allocation, and finally, minimize the bit error rate (BER) and packet error rate (PER). All these aspects fully correspond to the main aim of the authors of this book, that is, to show the reader how all the basic components of the wireless network should be completely integrated:

- the physical layer, based on multipath fading phenomena;
- signal processing, based on modulation techniques;
- protocols and accesses of multiuser servicing;
- antenna design layer, based on the performance of multibeam and phased-array antennas, and so on.

Of course, there are other components of each wireless network – the architecture and electronic circuits, based on different elements of the system hardware, such as RAKE detector, correlators, and filters, and so on. These aspects are out of scope of this special issue, and we refer the reader to the excellent books presented in [17–21].

References

1 Blaunstein, N. and Christodoulou, C. (2014). *Radio Propagation and Adaptive Antennas for Wireless Communication Networks – Terrestrial, Atmospheric and Ionospheric*, 2e. Hoboken, NJ: Wiley.

2 3GPP Technical Specification Group Radio Access Network (2010). Evolved Universal Terrestrial Radio Access (E-UTRA), Physical Layer Procedures (Release 9), 3GPP TS36.213 V9.3.0, June 2010.

3 3GPP Technical Specification Group Radio Access Network (2010). Evolved Universal Terrestrial Radio Access (E-UTRA), Further advancements for E-UTRA physical layer aspects (Release 9), 3GPPTS36.814V9.0.0, March 2010.

4 Ghaffar, R. and Knopp, R. (2011). Interference-aware receiver structure for Multi-User MIMO and LTE. *EURASIP J. Wirel. Commun. Netw.* 40: 24.

5 Li, Q., Li, G., Lee, W. et al. (2010). MIMO techniques in WiMAX and LTE: a future survey. *IEEE Commun. Mag.* 48 (5): 86–92.

6 Kusume, K., Dietl, G., Taoka, H., and Nagata, S. (2010). System level performance of downlink MU-MIMO transmission for 3GPP LTE-advanced. *Proceedings of the IEEE Vehicular Technology Conference*, Spring (VTC '05), Ottawa, Canada (September 2010), 5 pages.

7 Covavacs, I.Z., Ordonez, L.G., Navarro, M. et al. (2010). Toward a reconfigurable MIMO downlink air interface and radio resource management: the SURFACE concept. *IEEE Commun. Mag.* 48 (6): 22–29.

8 EU FP7 Project SAMURAI – Spectrum Aggression and Multi-User MIMO: Real-World Impact. http://www.ict-samurai.eu/page1001.en.htm.

9 3GPP TSG RAN WG1 #62 (2010). Way forward on transmission mode and DCI design for Rel-10 enhanced multiple antenna transmission, R1-105057, Madrid, Spain, August 2010.

10 3GPP TSG RAN WG1 #62 (2010). Way forward on 8 Tx Codebook for release 10 DL MIMO, R1-105011, Madrid, Spain, August 2010.

11 3GPP TR 36.942 V10.3.0 (2012-06), 3rd Generation Partnership Project (2012). Technical Specification Group Radio Access Network; Evolved Universal Terrestrial Radio Access (E-UTRA); Radio Frequency (RF) system scenarios (Release 10), June 2012.

12 3G Americas white paper (2010). 3GPP mobile broadband innovation path to 4G: release 9, Release 10 and beyond: HSPA+, SAE/LTE and LTE-advanced. http://www.4gamericas.org/documents/3GPP_Rel-9_Beyond%20Feb%202010.pdf (accessed 12 May 2020).

13 Duplicy, J., Badic, B., Balraj, R. et al. (2011). MU-MIMO in LTE systems. *EURASIP J. Wirel. Commun. Netw.* Article ID 496763, 13 pages. https://doi.org/10.1155/2011/496763.

14 3GPP TR 36.942 V8.4.0 (2012-06), 3rd Generation Partnership Project (2012). Technical Specification Group Radio Access Network; Evolved

Universal Terrestrial Radio Access (E-UTRA); Radio Frequency (RF) system scenarios (Release 8), June 2012.

15 Zhang, H., Prasad, N., and Rangarajan, S. (2011). MIMO Downlink Scheduling in LTE and LTE-Advanced Systems. *Tech. Rep.*NEC Labs America. http://www.nec-labs.com/honghai/TR/lte-scheduling.pdf (accessed 12 May 2020).

16 Blaunstein, N.Sh. and Sergeev, M.B.S. (2012). Integration of advanced LTE technology and MIMO network based on adaptive multi-beam antennas. In: *Internet of Things, Smart Spaces, and Next Generation Networking*(ed. S. Andreev, S. Balandin, and Y. Koucheryavy) (Proceedings of 12th International Conference on NEW2AN 2012 and 5th Conference ruSMART 2012, St. Petersburg, Russia (27–29 August 2012)), 164–173. Heidelberg, Dordrecht, London, New York: Springer-Verlag.

17 Molisch, A.F. (2007). *Wireless Communications*. Chichester: Wiley.

18 Rappaport, T.S. (1996). *Wireless Communications: Principles and Practice*, 2e in 2001. New York: Prentice Hall PTR.

19 Steele, R. and Hanzo, L. (1999). *Mobile Communications*, 2e. Chichester: Wiley.

20 Goodman, D.J. (1997). *Wireless Personal Communication Systems*. Reading, MA: Addison-Wesley.

21 Schiller, J. (2003). *Mobile Communications, Addison-Wesley Wireless Communications Series*, Reading, MA: Addison-Wesley 2e.

10

Satellite Communication Networks

10.1 Overview of Satellite Types

Communication by satellite is a man-made communication based on an artificial satellite that creates a communication channel between a source transmitter and a receiver at different locations on Earth. Communications satellites are used for various needs, such as television, telephone, radio, internet, and military applications.

The advantage of this technology is the great coverage area that cannot be achieved by any other mean. On the other hand, it is very complex and requires great resources. There are three main types of satellites, geostationary orbit (GEO), medium orbit (MEO), and low Earth orbit (LEO).

Geostationary satellites have a GEO, which is 36 000 km from the Earth's surface. It is called geostationary because in relation to the Earth it is stationary, since the satellite's orbital period is the same as the rotation rate of the Earth. Because the relation between the satellite and the Earth is constant, ground antennas do not have to track the satellite across the sky, they can be fixed to a point at the location in the sky where the satellite appears. This characteristic is a big advantage. MEO and LEO satellites are closer to Earth. MEO orbital altitudes range from 2000 to 36 000 km above Earth, and LEO satellites orbit between 160 and 2000 km above Earth.

MEO and LEO satellite's orbital period is greater than the rotation rate of the Earth. For that reason, they are not stationary and appear to be at different relation to the Earth at different time points. To a fixed position on Earth, those satellites seem as if they are crossing the sky. In order to establish continuous communication with these lower orbits, there is an obvious need for at least one satellite to be present in the sky with connectivity to any fixed point on Earth, for transmission of communication signals. This necessity requires a larger number of satellites. However, due to their relatively small distance to the Earth, their signals are stronger [1–4]. A comparison table between satellite orbits is given in Table 10.1.

Advanced Technologies and Wireless Networks Beyond 4G, First Edition.
Nathan Blaunstein and Yehuda Ben-Shimol.
© 2021 John Wiley & Sons, Inc. Published 2021 by John Wiley & Sons, Inc.

Table 10.1 Comparison between LEO, MEO, and GEO satellites

Characteristics	LEO	MEO	GEO
Altitude	160–2000 km	2000–36 000 km	35 786 km
Orbital period	100 min	6 h	Stationary
Number of satellites for global coverage	48–66	10–12	3
Space segment cost	Highest	Lowest	Medium
Satellite lifetime (years)	Imperceptible	Imperceptible	Long
Propagation delay call handover	Frequent	Infrequent	None

10.2 Signal Types in LSC Links

Propagation between a satellite and a mobile receiver can be classified as either nonshadowed, when the mobile or stationary subscriber has an unobstructed line-of-sight (LOS) path to the satellite, or shadowed, when the LOS path to the satellite is obstructed by either feature placed at the terrain, natural or man-made[1–14]. The nonshadowed signal received at the mobile receiver is composed of three signal components: direct, specular, and diffuse (see Figure 10.1). Propagation measurements indicate that a significant fraction of the total energy arrives at the receiver by way of a direct path. The remaining power is received by the specular ground-reflected path and the many random

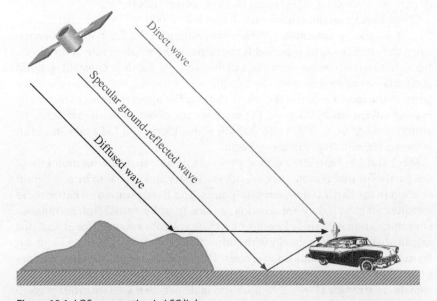

Figure 10.1 LOS propagation in LSC link

scattering paths that form the diffuse signal component. The shadowed signal occurs when the signal fade is caused primarily by scattering and absorption from obstacles such as both branches and foliage building roofs. Measured results indicate that shadowing is the most dominating factor determining slow signal fading. Its effect depends on the signal path length through the obstruction, type of obstruction, elevation angle, direction of traveler, and carrier frequency.

The shadowing is more severe at low elevation angles where the projected shadow of the obstacle is high. The effect of shadowing induced by vegetation is more complicated and depends on how frequently trees intercept, the path length through the trees, and the density of the branches and foliage. At frequencies lower than 1 GHz, trees are virtually transparent to the signal. For higher frequencies, trees are regarded as ideal edge refractors in order to estimate the amount of signal attenuation. The shadowed signal received at the mobile receiver is composed of two components: a shadowed direct component and a diffuse component. The shadowed direct component is generated when the LOS signal from the satellite passes through roadside vegetation and is attenuated and scattered by the leaves, branches, and limbs of the vegetation. The attenuation of the direct component depends on the path length through the vegetation. Scattering by the vegetation generates a random forward scattered field that interferes with the direct component causing it to fade and lose its phase coherency. Thus, the shadowed direct component can be modeled as the sum of an attenuation LOS signal and a random forward scattered field:

$$P_{\text{shadowed}} = \alpha P_{\text{direct}} + P_{\text{scattered}} \tag{10.1}$$

Here, α represents the attenuation factor of the direct component of the signal power, and $P_{\text{scattered}}$ is the scattered signal power from the vegetation. A typical shadowing attenuation from a building, bridge, or trees is on the order of 8–20 dB relative to the signal mean value.

The diffuse component results from various reflections from the surrounding terrain. This component varies randomly in amplitude and phase. Multipath propagation does not cause significant losses for land mobiles.

The shadowed diffuse component from vegetation is identical in form to the diffuse component for nonshadowed propagation. The diffuse component is assumed to be received randomly from all angular directions. Hence, the total shadowed signal is the sum of the shadowed direct component and the diffuse component:

$$P_{\text{shadowed}} = \alpha P_{\text{direct}} + P_{\text{scattered}} + P_{\text{diffuse}} \tag{10.2}$$

We can use expression (10.2) to calculate the total path loss within the land–satellite communication (LSC) link. For this purpose, we need to use the corresponding models, pure statistical or physical–statistical, based on some special experiments and numerous measurements.

10.3   Overview of Experimentally Approbated Models

LSC systems enable users of handle wireless phones, portable computers, or mobile phones to communicate with one another from any two points worldwide. Signal propagation for such systems has become an essential consideration. Path conditions may cause harmful impairments that severely corrupt the system availability and performance. Hence, propagation considerations are very important for successful operation. Most satellites employ fixed, not mobile, terminals as in LSC systems. While terrestrial land mobile systems are often able to exploit relatively strong multipath, it will be power limited and dependent on the LOS component. Satellite–mobile links operate with low signal margins, and obstructions due to overpasses and vegetation will cause outages and reduce communication quality.

Therefore, to design successful wireless LSC links, stationary or mobile, it is very important to predict all propagation phenomena occurring in such links, to give a physical explanation of the main parameters of the channel, such as path loss and slow and fast fading, and finally, to develop a link budget compared to the total noise at the outputs of the terminals of the channel. Moreover, in LSCs, we must divide the channel into three parts. The upper channel covers the ionospheric radio propagation, and the middle part covers hydrometeors (such as rain, snow, and smoke). The main effects in the link budget and the total path loss come from the bottom part of the land–satellite channel, where effects of the terrain profile, which cause shadowing (or slow fading), become more appropriate.

In the land subchannel, local shadowing effects, caused by multiple diffraction from numerous wedges and corners of obstructions, become predominant and can significantly corrupt sent information from the ground-based terminal to the satellite and conversely. Furthermore, in the LSC typical land built-up scenarios, the LOS path between the satellite and the land terminal (stationary or mobile) can be affected by multipath mechanisms arising from reflection on rough ground surfaces and wall surfaces, multiple scattering from trees, and obstacles.

As was mentioned by Saunders [2], in such very complicated environments, accounting high-speed satellite movements, it is very complicated to differentiate slow- and fast-fading effects, as was done for land communication links analysis; they must be accounted for together. Therefore, we try to show how to take into account effects of the terrain built-up profile and multiple diffraction and scattering effects for fading description and for link budget design within the LSC channel. All these elements will allow us to predict the fading in the channel.

We analyze two main concepts on how to account for the terrain effects on LSCs. The first is based on the statistical models, whereas another is based on the physical–statistical models. To unify these models and to use them together in our analysis, we assume that the radio signal is moving within a channel only between two states: *good* and *bad*, as will be shown in the Lutz model.

The statistical model, which we present here, is based on the transfer from bad to good states and vice versa. At the same time, the physical–statistical models are based on the classical aspects of radio propagation over the terrain and the statistical description of obstructions placed randomly on the rough terrain. As will be shown below, such models can also be adapted to use the Markov's chain, as a basic aspect of pure statistical models. So, despite the fact that some researchers separate statistical and physical–statistical models and show them separately, we show how to unify these approaches. Both statistical parameters of Markov's stochastic process, statistical distributions of the built-up terrain features, and propagation phenomena can be used to predict the radio coverage of the Earth's surface and satellite constellation for the mega-cell maps design.

10.3.1 Lutz Pure Statistical Model

This model gives the possibility to estimate the effect of fading via parameter A, which gives information on how much time fading affects information data passing the channel [15]. It is based on Markovian chain. The simple statistics of LOS is translated as "good," and non line of sight (NLOS) is translated as "bad" (see Figure 10.2). This dichromatic separation defines well the large difference between the shadowed and nonshadowed statistics in urban and suburban areas. The parameters for each state and the transition probabilities for evolution between states are empirically derived. In the good state, the signal is assumed to be Rician distributed with K-factor, which depends on the satellite elevation angle and the carrier frequency, so that the PDF of the signal amplitude is given by $P_{good} = P_{Rice}$. In the bad state, the fading statistics of the signal amplitude are assumed to be Rayleigh, with a mean power $S_0 = \sigma^2$, which varies with time. For that reason, the PDF of amplitude is specified as the conditional distribution $P_{Ray}(S|S_0)$, where S_0 varies according to a lognormal distribution $P_{LN}(S_0)$, representing the varying effects of shadowing with

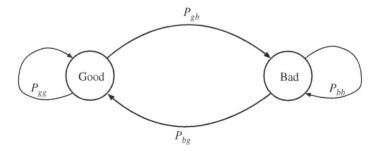

Figure 10.2 Markov model of the channel state

the NLOS component. Transitions between states are described by a first-order Markov chain. This is a state transition system in which the transition from one state to another depends only on the current state rather than on any more distant history of the system. The transition probabilities which summarize all the mentioned models based on Markov chain:

- Probability of transition from good state to good state – P_{gg}.
- Probability of transition from good state to bad state – P_{gb}.
- Probability of transition from bad state to bad state – P_{bb}.
- Probability of transition from bad state to good state – P_{bg}.

For a digital communication system, each state transition represents the transition of one symbol. The transition probabilities can then be found in terms of the mean number of symbol duration spent in each state [15].

$$P_{gb} = 1/D_g, \quad P_{bg} = 1/D_b \tag{10.3}$$

Where D_g is the mean number of symbol duration in the good state, and D_b is the mean number of symbol duration in the bad state. The time share of shadowing (the proportion of a symbol in the bad state) is

$$A = \frac{P_{gb}}{P_{gb} + P_{bg}} \tag{10.4}$$

The sum of the probabilities leading from any state must be equal to the sum of the unit; therefore,

$$P_{gb} + P_{gg} = 1, \quad P_{bg} + P_{bb} = 1 \tag{10.5}$$

This model will be used to compare with the Saunders–Evans physical–statistical model (see below).

10.3.2 Physical–Statistical Approach

In pure statistical models, the input data and computational effort are quite simple, as the model parameters are fitted to the measured data. Because of the lack of physical background, such models only apply to environments that are very close to the one they have been inferred from. On the contrary, pure deterministic physical models provide high accuracy, but they require actual analytical path profiles and time-consuming computations.

A combination of both approaches has been developed by the authors. The general method relates any channel simulation to the statistical distribution of physical parameters, such as building height, width and spacing, street width, or elevation and azimuth angles, of the satellite link. This approach is henceforth referred to as the *physical–statistical* approach [2, 14, 16].

As for physical models, the input knowledge consists of electromagnetic theory and a full physical understanding of the propagation processes. However, this knowledge is then used to analyze a statistical input data set, yielding a distribution of the output predictions. The output predictions are not linked to specific locations. Physical–statistical models therefore require only simple input data such as distribution parameters (e.g. mean building height and building height variance, as was done in Chapters 3–5 for land and atmospheric communication links). This model describes the geometry of mobile–satellite propagation in built-up areas and proposes statistical distributions of building heights, which are used in the subsequent analysis. We consider only two of them, which have been fully proved by numerous experiments for land–land and LSC links:

- a model of shadowing based on the two-state channel Lutz model;
- a multiparametric stochastic model.

10.3.2.1 Saunders–Evans physical–statistical model

The geometry of the situation is illustrated in Figure 10.3. It describes a situation where a mobile is located on a long straight street and a direct ray from the satellite to the mobile from an arbitrary direction.

The street is lined on both sides with buildings whose height varies randomly. In the presented model, the statistics of the building height in typical built-up areas will be used as input data. The PDFs that were selected to fit the data are the lognormal and Rayleigh distributions with unknown parameters of a mean

Figure 10.3 Geometry for mobile–satellite communication in built-up areas

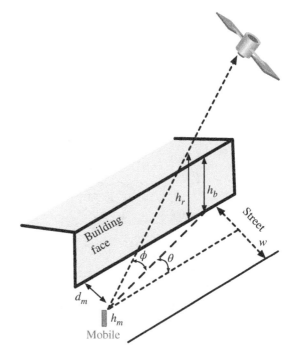

value, m, and standard deviation, σ^2. The PDF for the lognormal distribution is

$$p_b(h_b) = \frac{1}{\sqrt{2\pi}h_b\sigma_b} \exp\left(-\frac{\ln^2(h_b/m)}{2\sigma_b^2}\right) \tag{10.6}$$

The PDF for the Rayleigh distribution is

$$p_b(h_b) = \frac{h_b}{\sigma_b^2} \exp\left(-\frac{h_b^2}{2\sigma_b^2}\right) \tag{10.7}$$

Two appropriate parameters for these functions were found by measurements in the areas in which they were relevant to find the appropriate parameters for these functions in order to fit the data measurements as accurately as possible, the probability density function was found by minimizing the maximum difference between the two cumulative distribution functions.

The direct ray is judged to be shadowed when the building height h_b exceeds some threshold height h_T relative to the direct ray height h_r (see Figure 10.4). The shadowing probability, P_s, can then be expressed in terms of the probability density function of the building height, $P_b(h_b)$, as in Ref. [2]:

$$P_s = \Pr(h_b > h_T) = \int_{h_T}^{\infty} p_b(h_b)\, dh_b \tag{10.8}$$

The definition of h_T is obtained by considering shadowing to occur exactly when the direct ray is geometrically blocked by the building face (see Figure 10.3). Using a simple specific geometry (Figure 10.3) and general definitions of

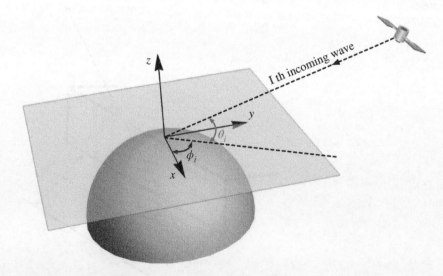

Figure 10.4 General geometry of land–satellite link

angles and (Figure 10.4), the following expression can be extracted for h_T:

$$h_T = h_r = \begin{cases} h_m + \dfrac{d_m \tan\phi}{\sin\theta}, & \text{for } 0 < \theta < \pi \\[2ex] h_m + \dfrac{(w - d_m)\tan\phi}{\sin\theta}, & \text{for } \pi < \theta < 2\pi \end{cases} \tag{10.9}$$

The shadowed model estimates the probability of shadowing for Lutz two-state model. The same Markov chain (as shown in Figure 10.2) is used, but parameters A, P_{bad}, and P_{good} are obtained from actual random distribution of the obstructions above the terrain. Parameter A is expressed as

$$A = \int_{h_1}^{h_2} p_b(h)dh \tag{10.10}$$

where h is the different heights of the obstacles, and h_1 and h_2 are the minimum and maximum heights of the built-up layer

$$p_b = \begin{cases} \text{lognormal + Rician} & \text{LOS conditions} \\ \text{lognormal + Rayleigh} & \text{NLOS conditions} \end{cases} \tag{10.11}$$

where lognormal PDF is pure NLOS shadowing, Rician PDF describes both the LOS and the multipath component, and Rayleigh's PDF describes the multipath component of the total signal, when the LOS component is absent.

$$p(S) = (1 - A)P_{good} + A \cdot P_{bad} \tag{10.12}$$

$$p(S) = (1 - A)P_{Rice}(S) + A \int_0^\infty P_{Ray}(S|S_0) \cdot P_{LN}(S_0)dS_0 \tag{10.13}$$

Then, we introduce the corresponding complementary cumulative distribution function (CCDF), which describes the signal stability, being the received signal with amplitude r that prevails upon the maximum accepted path loss (R) in the multipath channel, caused by fading phenomena. This can be presented in the following form:

$$\text{CCDF} = P_r(r > R) = \int_0^R p(S)dS \tag{10.14}$$

This model was the basic one used in satellite networks of the fourth generation. In the following section, we propose the same stochastic multiparametric model described in Chapters 3–5; some main formulas will be written again for the convenience to understand the matter by the reader.

10.3.2.2 Multiparametric stochastic model

The physical–statistical model, which is based on a deterministic distribution of the local built-up geometry, cannot strictly predict any situation when a satellite moving around the world has different elevation angles θ_i, with respect to a subscriber located at the ground surface, as shown in Figure 10.4. As a result, the radio path between the desired subscriber and the satellite crosses different overlay profiles of the buildings because of continuously changing elevation angle of the satellite during its rotation around the Earth, with respect to the ground-based subscriber antenna. To predict continuously the outage probability of shadowing in real time, a huge amount of data is needed regarding each building, geometry of each radio path between the desired user and the satellite during its rotation around the Earth, and finally, high-speed powerful computer.

Buildings' overlay profile The LSC link is very sensitive to the overlay profile of the buildings as shown in Figure 10.5, because during its movement around the Earth, depending on the elevation angle ϕ, the buildings' profile will be continuously changed, leading to different effects of shadowing in the current communication link (see Figure 10.5).

Because real profiles of urban environment are randomly distributed, the probability function $P_h(z)$, which describes the overlay profile of the buildings, can be presented in the following [16, 17]: for $n > 0, 0 < z < h_2$

$$P_b(z) = H(h_1 - z) + H(z - h_1) \cdot H(h_2 - z) \cdot \left[\frac{h_2 - z}{h_2 - h_1} \right]^n \tag{10.15}$$

where the function $H(x)$ is the Heaviside step function, which is equal to 1, if $x > 0$, and is equal to 0, if $x < 0$. For $n \gg 1$, $P_b(z)$ describes the case where buildings higher than h_1 (minimum level) very rarely exist. The case where all buildings have heights close to h_2 (maximum level of the built-up layer) is given by $n \ll 1$. For n close to zero, or n approaching infinity, most buildings have approximately the same level that equals h_2 or h_1, respectively. For $n = 1$, we have the case of the buildings' heights uniformly distributed in the range of

Figure 10.5 Change of the profile function $F(z_1, z_2)$ in the vertical plane during satellite fly

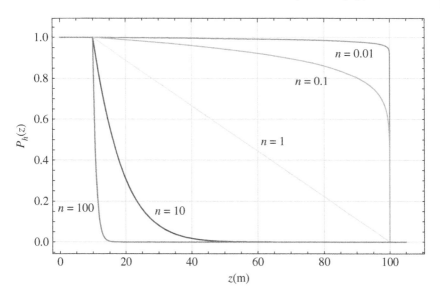

Figure 10.6 Distribution for a building present as a function of the subscriber's height and the terrain factor (n):for $h_1 = 10$ m and $h_2 = 100$ m

$h_1 - h_2$. The parameter n can be calculated as follows:

$$n = \frac{h_2 - \overline{h}}{\overline{h} - h_1} \tag{10.16}$$

$\overline{h}$ is given by an investigation of a topographic map of the terrain. In an urban environment, measuring all (or at least, most) of the building's heights, the average height can be found.

Finally, we can obtain the built-up profile between two terminal antennas for the case when the antenna height is above the rooftop level. According to the configuration of the land–satellite links, where $z_1 < h_2 < z_2$, we get

$$F(z_1, z_2) = H(h_1 - z_1) \left[h_1 - z_1 + \frac{h_2 - h_1}{n + 1} \right] +$$
$$H(z_1 - h_1)H(h_2 - z_1) \frac{(h_1 - z_1)(n + 1)}{(n + 1)(h_2 - h_1)^n} \tag{10.17}$$

Then, the CDF of the event that any subscriber located in the built-up layer is affected by obstructions due to shadowing effect can be presented as [18]

$$A = \frac{1}{z_2 - z_1} \int_{z_1}^{z_2} p_b(h)dh \tag{10.18}$$

$$\text{CDF}(z_1, z_2, n) = \frac{1}{z_2 - z_1} \int_{z_1}^{z_2} P_h(z)dz \equiv \frac{1}{z_2 - z_1} F(z_1, z_2) \tag{10.19}$$

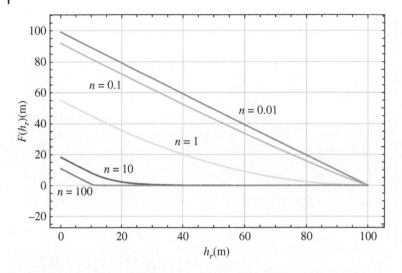

Figure 10.7 Distribution of $F(h_r)$ vs. h_r, as a function of the terrain factor n

To prove the assumptions presented above, we create a numerical experiment and compute the equations presented in the previous section. We use in our computations the following parameters of the built-up terrain and configuration of the land–satellite link: $h_2 = 100$ m, $h_1 = 10$ m, $z_1 = 3$ m, $z_2 = 100$ m, and $n \in \{0.01, 0.1, 1, 10, 100\}$.

Distribution for a building's overlay profile as a function of the subscriber's height $h_1 = 10$ m and the satellite virtual height of $h_2 = 100$ m (we take the height of the roof of the higher building which can intersect the LOS trace, see Figure 10.3) for various terrain factors $n = 0.01$ (sky-elevated buildings) to $n = 100$ (small buildings) is shown in Figure 10.6. It is seen that for $n \gg 1$, the profile limits to high-elevated buildings, and for $n \ll 1$, to small buildings. For $n = 1$, the number of high and small buildings is the same. This situation is illustrated by a straight line.

Figure 10.7 illustrates the changes of function of the buildings' profile with the change of the height of the receiving antenna. It is seen that with an increase of h_r, the influence of the built-up profile becomes weaker, the effect depending on the parameter of the terrain n.

A similar property is found for CCDF describing the diffraction effect on fading for various heights h_r and parameters n: the lower the parameter n, that is, the higher the building's heights, much more influence of diffraction of signal from building roofs is observed; this tendency is weak for small-building topography of built-up terrain (see Figure 10.8).

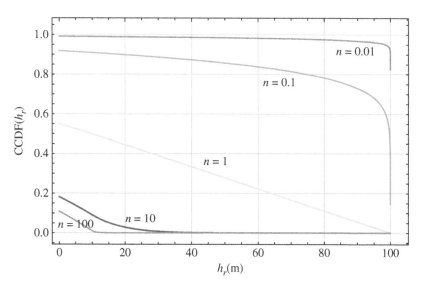

Figure 10.8 CCDF as a function of the height receiver antenna and the terrain factor (*n*)

10.4 Comparison Between Saunders–Evans and the Stochastic Multiparametric Model

We prove the accuracy of the stochastic model. To achieve this goal, we assist the Saunders–Evans model and the results obtained by Saunders experimentally. Based on the corresponding models and formulas presented above, we evaluate the *A* parameter, which is valuable in understanding the expected fading occurring in the land–satellite link. The *A* parameter will be compared to the parameter *A* computed by Saunders. Once shown identical, the Stochastic model will be proven correct.

According to Saunders–Evans model, above the Stockholm it was found that $A = 0.24$. Working with the topographic map of Stockholm (see Figure 10.9), it was found that the highest building in Stockholm is 120 m. We put it as h_2. The lowest building was taken $h_1 = 3$ m. By examining Stockholm's map, we found that the average height of the building was near 30 m. By examining the topographic map of Stockholm deeper, we obtained that $n = 3$, and that the parameter of fading $A = 0.25$ is very close to that found by Saunders and Evans.

It is clearly seen from the data computed according to the stochastic multiparametric model that it can be used as a stable and correct predictor of fading phenomena occurring in LSC links and proves the results obtained by Saunders and Evans experimentally above England and Scandinavian countries: with

Figure 10.9 2D topographic map of Stockholm. At the same manner, working with the topographic maps of small town, $n = 10$; medium city, $n = 1$; and large city, $n = 0.1$, we can predict the parameter of fading for these three typical cities (see Table 10.2)

increase of the height of buildings and decrease of their built-up terrain profile parameter n, the effect of fading becomes more significant and can achieve the probability of this phenomenon close to unit. In other words, the probability of fading and its relative duration (via parameter A) limits to unit. Thus, for small town $A = 0.091$, whereas for the large city $A = 0.909$, as well as for a medium city $A = 0.5$.

So, we can summarize that by the use of the fading parameter A, one can obtain information on the time fading affects data passing through the desired LSC channel. This parameter can be represented by the term AFD (average fade

Table 10.2 Building's overlay profile in three typical cities

City size	n	h_1,z_1(m)	h_2,z_2(m)	A
Small	10	3	15	0.091
Medium	1	3	35	0.5
Large	0.1	3	100	0.909

duration). Moreover, from AFD parameter one can estimate the bit-error-ratio (BER), which is a crucial parameter in any communication system. On the other hand, the parameter A shows for which percentage the channel is considered shadowed and for that reason from the data transmitted is not received. That is, the same meaning of BER, how many bits are not received in relation for the transmitted bits. For example, $A = 0.5$ will result in that 50% of the transmitted time will not exceed the LCR of the receiver (in average) and for that reason 50% from the transmitted bits would not be considered received.

10.5 Land–Satellite Networks – Current and Advanced Beyond 4G

In order to understand the operational characteristics of the land–satellite mobile communication systems, we start with the existing systems, following Refs. [3, 4, 16, 17]. Earlier mobile–satellite communication systems employed geostationary satellites with modest effective isotropic radiated power (EIRP), which restricts their use to mobile terminals with an antenna gain of approximately 8 dB (considerably more than can be achieved with a handheld unit, which might be only on the order of 2 dB). These terminals are most frequently mounted on vehicles, although fixed and transportable versions can also be provided.

10.5.1 Current Land–Satellite Networks

10.5.1.1 Inmarsat

The commercial use of satellites for mobile communications began with the *COMSAT/Marisat* system in 1976 [19]. Satellites operating at UHF ($\sim$ 800 MHz) and at L-band ($\sim$ 1600 MHz) were launched on 19 February 1976 and 9 June1976 into positions over the Atlantic and Pacific Oceans, respectively. The UHF capacity was utilized by the US Navy, while the L-band capacity was intended to inaugurate a commercial service for mariners. Shipboard terminals typically consisted of an above-deck 1 m diameter-specific antenna to remain locked on the satellite. To increase the efficiency of this

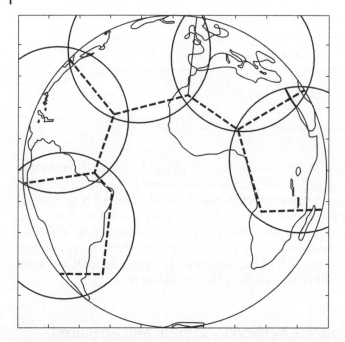

Figure 10.10 Mega-cell patterns of an *Inmarsat*-3 satellite system with five possible spot beams, four of which can be activated to provide coverage of land areas (according to [3])

system, as well as to create more channels, *Inmarsat* introduced three new services [3, 4]. In 1996, the first two of the five *Inmarsat*-3 satellites were launched. These satellites reuse the authorized frequencies in up to five spot beams, which can be selected for their coverage over land, as shown in Figure 10.10 according to Ref. [3].

10.5.1.2 North American MSAT system
The most advanced satellite system providing mobile–satellite communications over land is a system implemented by the United States and Canada that covers those two countries (including Alaska), Mexico, and the Caribbean using two essentially identical satellites. Early designs focused on a UHF system; however, a forward communication channel (FCC) proceeding concluded that any US system should operate at L-band. Moreover, as the mobile terminals could not readily discriminate between satellites placed in the longitude sector that covers North America, there could only be a single system. The contenders to provide this service were thus forced into a consortium, which became known as the American Mobile Satellite Corporation (AMSC) [3, 4].

In Canada, Telesat Mobile Inc. (TMI) was licensed to provide service, and these two companies, TMI and AMSC, agreed on a common set of specifications and an arrangement for sharing the available bandwidth. The ground

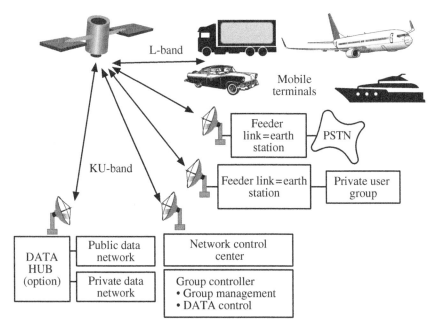

Figure 10.11 Arrangement of the ground segment for the AMSC/TMI system

segment for the AMSC/TMI system is depicted in Figure 10.11 according to [3].

This system differs from that employed in the *Inmarsat* system in that provision is made for separate Earth stations providing public switched telephone network (PSTN) services, private network services, and data services, all under the command of a network control center.

10.5.1.3   Australian mobile satellite system (OPTUS)
Australia has implemented a mobile–satellite service using an L-band transponder carried on the *Aussat-B* series spacecraft [3, 4]. Because of Australia's relatively high latitude, the elevation of the satellite, as seen from the more populated parts of Australia, was over 40°, somewhat simplifying the design of the mobile antenna. The services offered by this system include telephony using 4.8 kbps voice encoding in 5 kHz channels (similar to the *Inmarsat-M* system), as well as a data service at 2.4 kbps and fax at 4.8 kbps [3, 4]. Figure 10.12, according to Refs. [3, 4], presents the land coverage of *Aussat B-1* satellite by use of single spot beam.

10.5.1.4   Japanese N-Star mobile communications system
The Japanese NTT Mobile Communications Network (*NTT DoCoMo*) initiated mobile service in March 1996 using the *N-Star* satellites, offering both fixed and mobile services. These satellites provide mobile services to the Japanese islands

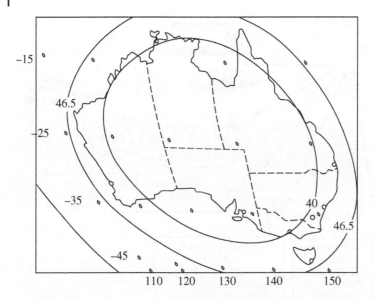

Figure 10.12 Coverage of Australia by the Aussat B-1 satellite (according to [3])

and surrounding waters via four 1200-km-diameter spot beams, as illustrated in Figure 10.13, according to Ref. [3, 4].

Three kinds of mobile terminals have been developed: portable, maritime, and car mounted. The maritime terminals use the satellite exclusively, and the car-mounted terminals provide access to both the terrestrial cellular network and the satellite system. The portable units are available in this dual-mode version also or in a satellite-only version. The system provides voice service using a 5.6- kbps codec, which results in a transmission rate of 14 kbps. Fax and data are sent at 4.8 kbps rates.

The channel spacing is at 12.5 kHz. Each of the two satellites is accessed by separate base stations, each of which has dual redundant equipment for high reliability [16].

10.5.1.5 Other mobile–satellite systems

Mexico also implemented a mobile–satellite service employing L-band transponders on each of the two Solidaridad satellites launched in 1994 [16].

India is planning a mobile–satellite service using an S-band (2 GHz) transponder placed on the INSAT 2 spacecraft [16]. This service will consist of voice, data, and fax using essentially the same parameters as the *Inmarsat*-M service.

During the 1990s of the past century, Russia has published information on the development of the global net of satellite called "GLONAS," which is planned to perform for a full coverage of territory of Russia by a family of small satellites to significantly increase the efficiency of service of stationary and mobile

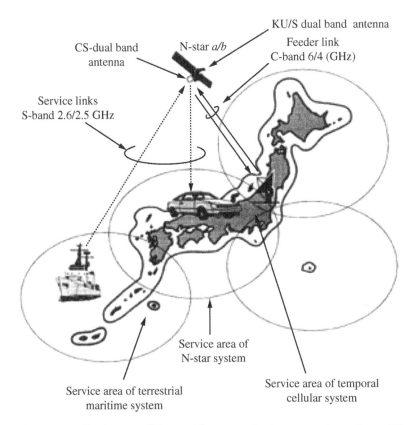

KU/S dual band antenna

Feeder link
C-band 6/4 (GHz)

N-star *a/b*

CS-dual band
antenna

Service links
S-band 2.6/2.5 GHz

Service area of
N-star system

Service area of terrestrial
maritime system

Service area of temporal
cellular system

Figure 10.13 The Japanese N-Star mobile communications system (according to [3])

subscribers located in various indoor/outdoor channels consisting multimedia (voice, video, and audio), digital data, and messages. During 2010/2011, several satellites of such types were successfully launched and started to operate in the ordinal working regime of global service.

10.5.2   Advanced Satellite Networks Performance

The proposed new global satellite PCS and their communication characteristics are presented in [3, 4, 16, 17]. Let us briefly discuss these new PCS [3].

10.5.2.1   Iridium
The design employs 66 satellites placed in circular polar orbits at 780 km altitude [3, 16]. The satellites are deployed into six orbital planes, with 11 satellites equally separated around each orbit. Satellites in adjacent planes are staggered with respect to each other to maximize their coverage at the equator, where a user may be required to access a satellite with elevation angle of less then 10 above the horizon.

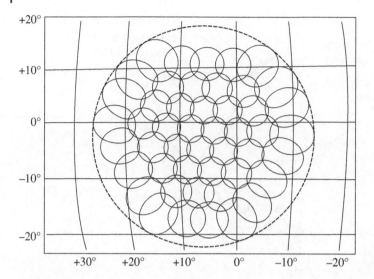

Figure 10.14 Service (L-band) spot beams formed by an *Iridium* satellite (according to [3, 16])

Users employ small handsets operating in frequency-division-multiplexed/time-division multiple access (FDM/TDMA) fashion to access the satellite at L-band. Eight users share 45 ms transmit and 45 ms receive frames in channels that have a bandwidth of 31.5 kHz and are spaced 41.67 kHz. That is, users are synchronized so that they transmit and receive at the same time using timewindows alternately. This approach is necessary because three phased-array adaptive antennas are used for both transmitting and receiving information. Figure 10.14 shows the 48 spot beams formed at L-band projected onto the Earth at the equator.

The satellite cross-links operate at 23 GHz, whereas the links to the gateway Earth stations are at 20 GHz. The use of cross-links greatly complicates the design of the system but allows global service to be provided with a small number (11 are planned) of gateway Earth stations. To properly route traffic, each satellite must carry a set of stored routing tables from which new routing instructions are called every 2.5 min.

The cross-links to the satellites in the adjacent orbital planes have constantly changing time delays and antenna pointing requirements. To mitigate this problem, a circular polar orbit (actually an inclination of 86.5°) was chosen. Even so, these cross-links are dropped above 68° latitude, as the angular rates for the tracking antennas become high, and little traffic is expected at these latitudes. Each satellite is capable of handling as many as 1100 simultaneous calls.

Services to be provided include voice (at 2.4 and 4.8 kbps encoding), data at 2.4 kbps, and high-penetration paging, which affords 11 dB more power than the regular signal. The design, however, already provides a link fade margin (16 dB) that is higher than any of the competing systems.

One of the complicating aspects of the *Iridium* system is the need to hand off a subscriber from beam to beam as a satellite flies by. As a typical satellite pass takes less than 9 min, and the average international call duration is about 7 min, there is also a need to hand off some calls to the next satellite to appear above the horizon. This will be in one of the adjacent orbits, and hence in a somewhat different direction from the first, raising the possibility of the calls being dropped if buildings block the view.

Other systems, such as *Globalstar*, attempt to exploit dual-satellite visibility as a means of mitigating shadowing effects and claim that this is preferable to designing for high link margins.

10.5.2.2 Globalstar

The *Globalstar* system has 48 satellites organized in eight planes of six satellites each (see Figure 10.15). The satellite orbits are circular, at 1414 km and 52° inclination. The *Globalstar* satellite is simple. Each satellite consists of an antenna, a trapezoidal structure, two solar arrays, and a magnetometer. As was

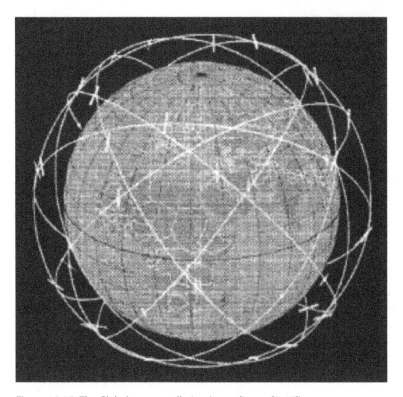

Figure 10.15 The *Globalstar* constellation (according to [3, 17])

Globalstar Gateway

☐ Extended Globalstar Service Area
☐ Extended Globalstar Service Area
   (Customers may have single satellite coverage and experience a weaker signal)
☐ Bring Globalstart Service Area
   (Customers may experience weak or sporadic service)
☐ Globalstar Service Area currently Available to North American Roamers

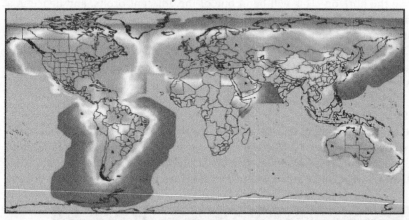

Figure 10.16 Up-to-date coverage map of *Globalstar* (according to *Globalstar* website)

shown in [3, 16], the use of an inclined orbit concentrates the available satellite capacity at lower latitudes where the largest populations exist.

Moreover, a little or no coverage is provided beyond 70° latitude (see Figure 10.16). Like the *Iridium* satellites, the *Globalstar* spacecrafts are three-axis stabilized, with a mission life of 7.5 years.

As the *Globalstar* system does not employ satellite cross-links, a subscriber can gain access to the system only when a satellite in view can also be seen by a gateway Earth station. Typically, this means that service areas are within 1000 miles of each gateway Earth station. To achieve truly global coverage would require the construction of more than 200 Earth stations, which seems unlikely to happen.

Moreover, in contrast to *Iridium*, each *Globalstar* satellite covers a comparable area of the Earth's surface with only 16 spot beams. Thus, *Globalstar* is more likely to serve national roamers than international business travelers. This, together with the sharing of the receive channels onboard the satellite by many more users, reduces the available link margins to about 3–6 dB, although for a small number of users, this can be increased to 11 dB. Access to and from the satellite is at L and S bands, respectively, utilizing code-division multiple access (CDMA) in channels that are 1.25 MHz in bandwidth. Voice is encoded at rates of 1–9 kbps, depending on the speaker activity. A problem for this satellite system is that while frequency reuse can be employed at L and S bands, the feeder links must occupy the full band of all of the signals that can be transmitted through the satellite. Therefore, securing an adequate feeder link allocation becomes almost as critical as an L- and S-band allocation.

10.5.2.3 ICO-global

ICO-Global has chosen an intermediate circular orbit for its system (10 355 km altitude), with 10 satellites arranged five in each of two inclined circular orbits [16]. The inclination of the orbits is 45°, making it the lowest of the systems described. This reduces the coverage at high latitudes but allows for the smallest number of satellites. Actually, 12 satellites are to be launched in order to provide a spare in each orbital plane.

To improve the link fade margins on the ICO satellites, *Inmarsat* chose a satellite design that employs 163 spot beams (see Figure 10.17, according to [16]). Therefore, to access a given spot beam, the gateway Earth station must transmit at a particular frequency. Given that a large digital signal processor is required onboard the ICO satellites, it is expected to greatly simplify the checkout and calibration of the adaptive antenna systems. In principle, this arrangement would also permit the beams to be steered to increase the amount of time a given subscriber remains in a given spot beam.

ICO differs from both *Iridium* and *Globalstar* in that a true TDMA scheme has been adopted for the service links, with six subscribers multiplexed into channels 25.2 kHz in width at a bit rate of 36 kbps. A disadvantage of this access scheme is that a soft handoff (e.g. from beam to beam) is not automatic, and it is more difficult to exploit dual-satellite visibility. One method being considered would be to send a burst via an alternate satellite (say) every fifth burst. By noting the strengths of the regular and alternate bursts, the subscriber terminal could determine which satellite presently affords the best path to the gateway Earth station and could adjust its own burst time and frequency to select that

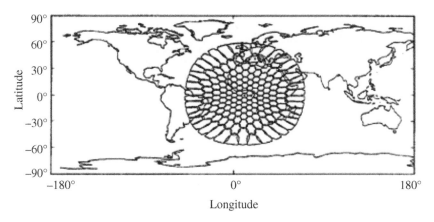

Figure 10.17 Coverage provided by the 163 spot beams of an ICO satellite over the equator (according to [16, 17])

satellite. For more coverage on other existing land–satellite PCS, the reader is referred to the original works [16, 17].

10.5.2.4 European Inmarsat BGAN

The *Inmarsat* satellites network described in the previous paragraph was also used for services of any subscriber located in disaster areas where it is impossible to use the ground-based networks, such as GSM infrastructure. To do that, an advanced *Inmarsat* satellite solution was arranged on the basis of Broadband Global Area Network (BGAN) technology (see more detailed information in Ref. [17]). The project that is based on such a technology was founded by the European Commission and is called the Wireless Infrastructure over Satellite for Emergency Communications (*WISECOM*) [17]. Thus, the *WISECOM* includes developing a complete solution that can be rapidly deployed immediately after the disaster, within the first 24 h, replacing the traditional use of satellite phones. In 2007, two operating satellites with the radio coverage as presented in Figure 10.18 (there are two satellites in the middle of the map) were launched; the third satellite was launched in 2008.

Such a configuration allows using GSM technology over BGAN system architecture. Moreover, the satellite terminal antennas provide GSM communication coverage in disaster areas, where existing ground-based communication infrastructure has been destroyed or overloaded. Finally, the *Inmarsat* BGAN system, using GSM backhauling over the corresponding satellite, transmits GSM signaling and data traffic to the core GSM network infrastructure in the specific geographic disaster-safe region at the Earth. What is interesting to note that such an advanced system is successfully operated for the regularly used modern (3.5G and 4G) wireless communication networks, such as Wi-Fi and WiMAX, operating with e-mail and voice over IP data

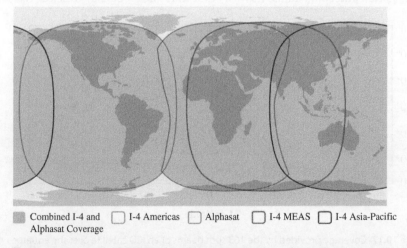

Combined I-4 and Alphasat Coverage   I-4 Americas   Alphasat   I-4 MEAS   I-4 Asia-Pacific

Figure 10.18 Inmarsat BGAN world radio coverage (extracted according to [16, 17])

transport [16, 17]. Thus, *WISECOM* restores local GSM or 3G infrastructures, allowing normal mobile phones to be used and enables wireless standard data access (e.g. Wi-Fi or WiMAX). Integration of Wi-Fi and GSM is the guarantee of a high-level quality of service for critical voice services when the traffic is mixed with less-critical internet traffic. The proposed system uses lightweight and rapidly deployable technologies, the so-called WISECOM Access Terminal (WAT), which can be carried by one person onboard a flight and be deployed within minutes. For more details, we refer the reader to the corresponding Refs. [16, 17].

10.5.2.5 Advanced GSM–satellite network

The same solution of how to use a quick and cost-effective GSM link over satellite network using a new interface over digital video broadcasting (called DVB-S2) was proposed in Refs. [16, 17]. The authors have performed a new interface (called "Abis") to evaluate the good performance of the land–satellite link in conditions of high signal delay and its power loss within such a channel. We refer the reader to these works since a lot of aspects described there are out of scope of our book, namely, new advanced protocols, packets transport, and the lossless voice over IP solutions.

10.5.3 Operational Parameters Prediction in Advanced Land–Satellite Networks

Prediction of radio and cellular maps for full coverage of the Earth's surface, using determined constellations of satellites for specific networks, CEO, MEO or LEO, is based on rigorous statistical and physical–statistical propagation models, which take into account the average path loss along all three "sublinks": land, atmospheric, and ionospheric. These effects can be postulated as worst (or bad) or convenient (or good) with intermediate variants "good–bad" and "bad–good," obtained from numerous measurements carried out in these three "sublinks." It was shown in Refs. [16, 17] that for the frequencies of interest operated in land–satellite links, the effects of these two "subchannels" (tropospheric and ionospheric) on total path loss and fading are not so significant (see the corresponding tables in Chapter 3). More essential fading effects are observed in the land communication "subchannel."

To show the reader how to predict the land-link total path loss and to obtain "mega-cell" radio coverage using determined satellite constellations, corresponding to more applicable satellite networks, different mathematical tools were developed [2–5, 15–18, 20, 21]. The proposed tool in Section 10.3, a stochastic multiparametric physical statistical approach, showed the efficiency of predicting fade margins and the probability of fading. At the same time, another planning tool was developed, which combined the two models Loo's pure statistical model, which is close to Lutz and Abdi models, and three-stated Markov model (see description of these models in [2, 16, 17]). For this purpose, the authors, analyzing different satellite networks, introduced the PDFs, $P_a(r)$, $P_b(r)$, and $P_c(r)$, for LOS, multipath, and shadow effects description, respectively. Here, a, b, and c correspond to the transaction from bad to good situations within a wireless link, depending on various environment phenomena. Then, for each situation within the channel, the authors in [2–5, 16, 17] took the corresponding PDFs, $P(a)$, $P(b)$, and $P(c)$ (the lognormal, Rayleigh, and Rician, respectively). Next, the following cumulative formula for determining the probability of fading phenomena within the satellite link was found:

$$P_{\text{total}} = P(a)P_a(r) + P(b)P_b(r) + P(c)P_c(r) \tag{10.20}$$

Below, we present the different coverage maps based on the planning tool developed in [2–5, 16, 17] on the basis of the cumulative formula (10.20) in order to understand the differences between satellite networks and their effect on the network's *footprint*. Thus, in Figure 10.19, the cellular map of *Iridium* network [3, 16] with a uniform radio coverage of the Earth's surface is presented.

Figure 10.20 illustrates the cellular map for *Globalstar* network [16, 17]. The corresponding constellation covers most of the populated area of the Earth. However, the polar areas of the Earth cannot be covered by this system.

Another example of how to create the coverage radio map is shown in Figure 10.21 for ICO network according to computations made in Ref. [17]. It is seen that a MEO satellite constellation covers the Earth with only 12 satellites with great overlapping, which is caused by their relatively high altitude.

Now we present some outputs from our calculations of the general physical–statistical model described by another cumulative formula (10.20), proposed by the authors on the basis of results obtained in [16, 17], to

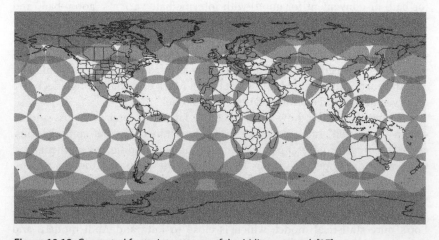

Figure 10.19 Computed footprint patterns of the *Iridium* network [17]

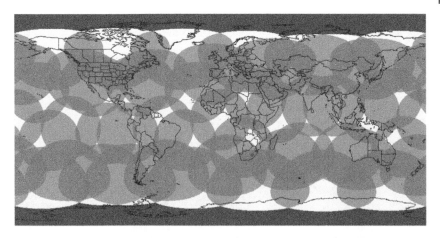

Figure 10.20 Computed footprint patterns of the *Globalstar* network [16, 17]

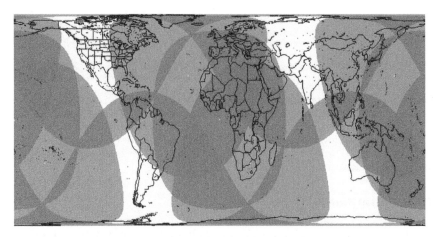

Figure 10.21 Computed footprint patterns of the *ICO* network [17]

demonstrate the actions of their own simulation tool and its ability to be adapted to different cellular networks. The output parameters that can be computed by the proposed tool are probability of fading, path loss, link budget, radio and cell coverage, LCR, AFD, and BER.

As an example, we varied the types of the satellite networks and investigated the outage probability of shadowing (fading) vs. the maximum accepted path loss for different satellite networks. Results are presented in Figure 10.22a–d, where one network is virtually created by the authors.

The difference between the networks is seen in the x-axis. It is clear that the differences in results are due to the varying satellite altitudes and the downlink operational frequency.

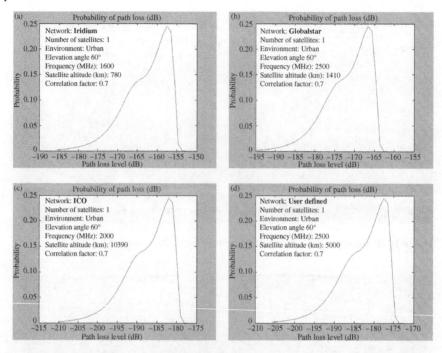

Figure 10.22 The simulated output of probability to obtain a smaller acceptable path loss for varying global networks based on operational parameters of: (a) Iridium, (b) GlobStar, (c) ICO, and (d) virtually created by the authors. The simulated dependences of probability to obtain the total path loss have similar "shape", but their maximum lies at different path loss ranges, differ on 10–20 dB from each other due to the varying satellite altitudes and downlink operational frequency of each individual network

10.6 Summary

From the presented results, we can conclude that the pure statistical model is the weakest model with respect to the physical–statistical models in predicting fading phenomena and link budget in LSC links. Also, in almost all simulations, the physical–statistical model based on built-up profiles described by the corresponding CDFs in expressions (10.12)–(10.14), and (10.19), following the multiparametric model, is the best fit to measurements with respect to pure statistical models. Furthermore, the difference between the two physical–statistical approaches, the Saunders and Evans [2, 3] and the multiparametric [21], on how to describe the overlay profile of the buildings is small, but the multiparametric model is much simpler to use and implement. The reason for this is the fact that for the physical–statistical model, we need exact knowledge of the distribution of the heights of the buildings in each city. This information is sometimes difficult to measure at every land site. On the contrary, for the multiparametric model, we need to know only the heights of the smallest and the highest building in the city and also the average height of buildings in every land site. From these values we could easily find the "relief" parameter n, using Eq. (10.17).

Both the Saunders and Evans physical–statistical and the multiparametric stochastic approaches are more accurate models, which can be used in predicting fading phenomena and link budgets, both for personal and mobile land–satelliteradio communication links.

A simulation tool, based on the combination of pure statistical and physical–statistical model, was shown for the link budget performance and for the outage probability of path loss prediction for more applicable satellite networks based on radio propagation characteristics within each communication channel.

The simulation tool is designed to be used both by LMS designers and customers. The designers can determine, using the simulator, the best satellite constellation and channel characteristics for each satellite system, in terms of the desired performance and cost. The customers can use the tool to determine which system is best suited for their needs, in their specific location in the world.

We can outline some practical conclusions and remarks, which emphasize the advantages in LMS systems. They are:

- a high elevation angle between the satellite and the LMS customer increases the link quality significantly. The main advantage in using the LMS system is emphasized dramatically in urban environments, where one satellite in a high-elevation angle eliminates the use of dozens of base stations;
- LEO satellites provide better link quality than MEO satellites, but more satellites are required to fully cover the Earth's surface. The main obstacle in designing an LMS system is determining the best-suited satellite constellation for potential customers;
- A system designed for a personal stationary subscriber will be much different from a system designed for moving subscribers. The proposed tools, described by (10.12)–(10.14) or (10.19) with (10.20), give the designer the option to set its own user-defined system and check if it suits its customers' needs.

References

1 Farserotu, J. and Prasad, R. (2002). *IP/ATM Mobile Satellite Networks*. Boston, MA and London: Artech House.
2 Saunders, S.R. (2001). *Antennas and Propagation for Wireless Communication Systems*, 2e. Chichester: Wiley.
3 Evans, J.V. (1998). Satellite systems for personal communication. *Proc. IEEE* 86 (7):1325–1341.

4 Wu, W.W. (1997). Satellite communication. *Proc. IEEE* 85 (6):998–1010.

5 Saunders, S.R. (1999). *Antennas and Propagation for Wireless Communication Systems*. New York: Wiley.

6 Bertoni, H.L. (2000). *Radio Propagation for Modern Wireless Systems*. Upper Saddle River, New Jersey: Prentice Hall PTR.

7 Blaunstein, N. (2004). *Wireless Communication Systems, Handbook of Engineering Electromagnetics*, Chapter 12 (ed. B. Rajeev). New York: Marcel Dekker.

8 Rappaport, T.S. (1996). *Wireless Communications*. New York: Prentice Hall PTR.

9 Barts, R.M. and Stutzmn, W.I. (1992). Modeling and simulation of mobile satellite propagation. *IEEE Trans. Anten. Propag.* 40 (4):375–385.

10 Jakes, W.C. (1974). *Microwave Mobile Communications*. New York: Wiley.

11 Steele, R. (1992). *Mobile Radio Communication*. New York: IEEE Press.

12 Stuber, G.L. (1996). *Principles of Mobile Communications*. Boston, MA and London: Kluwer Academic Publishers.

13 Lee, W.Y.C. (1989). *Mobile Cellular Telecommunications Systems*. New York: McGraw Hill.

14 MolischA.F. (2005). *Wireless Communications*. Chichester: IEEE Press.

15 Lutz, E., Cygan, D., Dippold, M. et al. (1991). The land mobile satellite communication channel-recording, statistics and channel model. *IEEE Trans. Veh. Technol.* 40 (2):375–385.

16 Blaunstein, N. and Christodoulou, Ch. (2007). *Radio Propagation and Adaptive Antennas for Wireless Communication Links*, 1e. Hoboken, New Jersey: Wiley InterScience.

17 Blaunstein, N. and Christodoulou, Ch. (2014). *Radio Propagation and Adaptive Antennas for Wireless Communication Networks*, 2e. Hoboken, New Jersey: Wiley InterScience.

18 Blaunstein, N. and Levin, M. (1996). VHF/UHF wave attenuation in a city with regularly spaced buildings. *Radio Sci.* 31 (2):313–323.

19 Blaunstein, N., Katz, D., Censor, D. et al. (2001). Prediction of loss characteristics in built-up areas with various buildings' overlay profiles. *IEEE Anten. Propag. Mag.* 43 (6):181–191.

20 Blaunstein, N., Cohen, Y., and Hayakawa, M. (2010). Prediction of fading phenomena in land-satellite communication links. *Radio Sci.* 45: 113–119, RS6005. https://doi.org/10.1029/2010RS004352.

21 Blaunstein, N. (2000). Distribution of angle-of-arrival and delay from array of building placed onrough terrain for various elevation of base station antenna. *J. Commun. Netw.* 2 (4):305–316.

Index

Advanced Technologies and Wireless Networks Beyond 4G, First Edition.
Nathan Blaunstein and Yehuda Ben-Shimol.
© 2021 John Wiley & Sons, Inc. Published 2021 by John Wiley & Sons, Inc.

Printed and bound by CPI Group (UK) Ltd, Croydon, CR0 4YY